Studies in Computational Intelligence

Volume 859

The series "Studies in Computational Intelligence" (SCI) publishes new developments and advances in the various areas of computational intelligence—quickly and with a high quality. The intent is to cover the theory, applications, and design methods of computational intelligence, as embedded in the fields of engineering, computer science, physics and life sciences, as well as the methodologies behind them. The series contains monographs, lecture notes and edited volumes in computational intelligence spanning the areas of neural networks, connectionist systems, genetic algorithms, evolutionary computation, artificial intelligence, cellular automata, self-organizing systems, soft computing, fuzzy systems, and hybrid intelligent systems. Of particular value to both the contributors and the readership are the short publication timeframe and the world-wide distribution, which enable both wide and rapid dissemination of research output.

The books of this series are submitted to indexing to Web of Science, EI-Compendex, DBLP, SCOPUS, Google Scholar and Springerlink.

More information about this series at http://www.springer.com/series/7092

Dylan Seychell · Alexiei Dingli
Editors

Rediscovering Heritage Through Technology

A Collection of Innovative Research Case Studies That Are Reworking The Way We Experience Heritage

Springer

Editors
Dylan Seychell
Faculty of ICT, Department
of Artificial Intelligence
University of Malta
Msida, Malta

Alexiei Dingli
Faculty of ICT, Department
of Artificial Intelligence
University of Malta
Msida, Malta

ISSN 1860-949X ISSN 1860-9503 (electronic)
Studies in Computational Intelligence
ISBN 978-3-030-36106-8 ISBN 978-3-030-36107-5 (eBook)
https://doi.org/10.1007/978-3-030-36107-5

This Springer imprint is published by the registered company Springer Nature Switzerland AG
The registered company address is: Gewerbestrasse 11, 6330 Cham, Switzerland

To my wife, Christine, with whom I travel the world. To all the fantastic people we met while travelling who served as an inspiration for this book. May technology help the rest of the world appreciate how awesome they and their heritage are.

Dylan Seychell

To the creator who, in his goodness, chose an awesome spacetime location for me. A space flowing with amazing people like my dear parents, my loving wife and our wonderful children; Ben, Jake and Faye. A lifetime full of marvellous memories and an exciting future ahead.

Alexiei Dingli

Preface

Art, in all its forms, has always acted as an indicator and benchmark of social progress. Throughout the years, it was a significant agent in paradigm shifts that brought around new ways of thinking and perspectives. It was the first indication of essential human communication, eventually evolving into a method of communicating piety. Its absence in medieval Europe also labelled the period as the dark ages. This darkness was suddenly lit and conquered upon by the Renaissance, where art started to blend with scientific principles, gently leading to eternal masterpieces. At this stage, the role of science was to make art more realistic and appealing, significantly improving its quality, relevance and value.

With the proliferation of computers, science asserted its role as a medium for creating modern pieces of art while refining and redefining what humanity achieved until that point. Computers are more powerful than just a tool or medium to assist in the creation of artistic work. They are helping all stakeholders to interact differently with art. Besides being a tool for artists, computers are helping historians to discover new features and attributes while aiding conservators to document art and carefully plan interventions. Above all, they are there for the rest of society to get closer to heritage through different dimensions. Technology is now providing us with a variety of tools to explore, experience and interact with art in ways we never imagined while bringing art closer to those who might not have considered it to be in their interest.

Artificial Intelligence, mainly through the recent developments in the evolution of deep learning techniques, is bringing a never-ending list of paradigm shifts to the way humanity operates. We have seen how AI is changing our future in novel ways that are now available at our fingertips. We have experienced changes in the way we choose music, explore cities in which we have never been before and also finding tailored content that we never thought could even exist.

Pause for a few seconds and think about how all this can help us even get a better understanding of our past by giving it a future through technology. The power of AI spans towards different aspects of technology and improves other approaches. Gamification, the process of applying game mechanics to situations other than games, is an innovative way of improving engagement. The exploration of heritage is confirmed to improve when gamification is adequately applied. Moreover, the use of novel visualisation techniques, such as Virtual Reality and Augmented Reality, provides a different way of exploring content. As we explored in our previous book, "The New Digital Natives—Cutting the Chord" in 2015, modern generations of users prefer innovative approaches to consume content that is otherwise perceived as monotonous.

When we designed the editorial concept of this book, we wanted to bring together our passion for travel, culture and heritage with our professional experience in Artificial Intelligence and technology. This book is envisaged to be a tool in bridging these two aspects that are so important for humanity. One of the key motivations was the variety of outstanding work by fellow researchers exploring topics that help build this bridge. As the world experiences isolating situations such as the COVID-19 pandemic, the importance of using technology for the rediscovery of heritage becomes ever more relevant. Through our academic ventures, we met passionate researchers who invest their time and dedicate their efforts to creating a better tomorrow for heritage through the use of technology. The variety of topics presented in this book is witness to these efforts, and we would like to take this opportunity to thank every researcher who contributed to this book.

The contributions of this book come from several highly respected international universities and research institutes. These are the University of Malta, Saint Martin's Institute of Higher Education, Centro Nazionale della Ricerca, Universita degli Studi di Palermo, CNIT Consortium, Universita degli Studi di Modena e Reggio Emilia, Universita degli Studi Roma Tre, Istituto Tecnico Tecnologico "Buonarroti", Universita Cattolica del Sacro Cuore and Danube University Krems.

This book is a carefully curated collection of chapters that explores a variety of ways in which technology is enabling us to take this new paradigm shift in experiencing heritage. The vital technological aspects discussed above guided the structure of this book. The book is organised into four parts. The first part presents work that employs AI and machine learning techniques to provide an enhanced interpretation of heritage by using Computer Vision and Natural Language Processing techniques. The second part explores the possibilities of employing gamification to improve the user experience in this domain. The last two parts present a collection of work that uses either Virtual Reality or Augmented Reality to present content. These case studies that are also supported by a detailed overview of UX design will undoubtedly serve as an inspiration and guide for future projects.

There is no reason why the future of design and implementation of heritage-related experiences should happen without the engagement of technology. The use of technology to rediscover heritage will guarantee the future appreciation of the past that brought humanity to what it is today.

Msida, Malta
July 2019

Dylan Seychell
Alexiei Dingli

Contents

Virtual Reality

Augmented Reality

About the Editors

Dylan Seychell is a resident academic at the University of Malta within the Department of Artificial Intelligence specialising in Computer Vision and User Experience Design, and a visiting academic at St Martin's Institute of Higher Education. He published a number of international peer-reviewed papers on the application of AI and Computer Vision in a variety of domains. He was awarded a number of international awards for his work such as the Gold Seal for e-Excellence at CeBit, the first prize by the European Space Agency Satellite Navigation Competition in 2010 and runner up in 2017. In 2015, Dylan was selected to lead Malta's Google Developers Group and served as a member of the Maltese Government national AI task-force tasked with the drawing up for a national strategy for Artificial Intelligence. Dylan founded successful startups in tourism and technology. He also serves as a technology advisor on a number of high-profile heritage projects.

Alexiei Dingli is a Professor of Artificial Intelligence (AI) and Head of the Department of AI at the University of Malta. He has been conducting research and working in the field of AI for the past two decades. His work was rated World Class by international experts and won various prizes including the Semantic Web Challenge, the first prize by the European Space Agency, the e-Excellence Gold Seal award, the first prize in the Malta Innovation Awards, the World Intellectual Property Organisation (WIPO) award for Creativity and the first prize of the Energy Globe award by the UN amongst others. He has published several peer-reviewed papers and books in the field. He is also involved in various AI projects with local and international organisations. The Government of Malta appointed him on the Malta. AI task-force aimed at making Malta one of the top AI countries in the world.

List of Figures

An Overview of Handwritten Character Recognition Systems for Historical Documents

Win the City Game

A Case Study into the User Experience of an Application of Virtual Reality at the Saint Paul's Catacombs, Malta

Techniques of Filming and Audio Recording in 360-Degree Ambiences

Augmented Reality in Cultural Heritage: Designing for Mobile AR User Experiences

List of Tables

Artificial Intelligence

AI is everywhere. It is not that big, scary thing in the future. AI is here with us.
Fei-Fei Li

An Overview of Handwritten Character Recognition Systems for Historical Documents

Mark Bugeja, Alexiei Dingli and Dylan Seychell

Abstract Analysing Handwritten Documents is a challenging task. This particular area cannot always come up with general solutions, given that most handwritten manuscripts contain unique characteristics that describe how the document was written, which include different handwritings. These challenges in transcribing different handwriting styles are due to various scribes contributing to the transcription of the text and degradation of the script. In this chapter, an overview of different techniques used in handwritten text recognition systems is presented. The approaches and algorithms can be adopted for different document types irrespective of the state of the scanned documents. Moreover, two different general approaches to handwritten character recognition are shown. The first approach goes through a fairly standard process to normalise, segment and recognise characters. The other approach is a segmentation free approach that uses neural networks for both segmentation and recognition.

1 Introduction

In today's world, printed media such as newspapers and magazines are being bought and consumed less [13]. On the other hand, society is moving towards more digitised formats that can be viewed using smartphone devices, tablets and ebook readers. This new way of consuming printed media is easier for consumers to digest and allows for other tools to be used, such as word search. Research in the area of optical character readers (OCR) and handwritten character readers (HCR) has been going on for decades [10, 39]. Advances in these research areas were made in order to

M. Bugeja · A. Dingli (✉) · D. Seychell
University of Malta, Msida, Malta
e-mail: alexiei.dingli@um.edu.mt

M. Bugeja
e-mail: mark.bugeja@um.edu.mt

D. Seychell
e-mail: dylan.seychell@um.edu.mt

D. Seychell and A. Dingli (eds.), *Rediscovering Heritage Through Technology*, Studies in Computational Intelligence 859,
https://doi.org/10.1007/978-3-030-36107-5_1

facilitate tasks such as document retrieval [3], library categorisation [31] as well as more practical field applications such as those used by post offices to automatically read addresses from a handwritten envelope [23]. OCR and HCR systems convert scanned versions of printed or handwritten characters into the digitised text. This process is also combined with other techniques that stitch the characters together to form words and lines, to output a digitised version of the printed content. To clarify, digitisation per se is the process of converting a document into a format that the computer can understand. Scanning a page is also digitising the page. OCR and HCR systems go a step further by converting the components of the text (characters) into ASCII representation, a format that is understood by machines. The documents that can be converted range from handwritten manuscripts to hard copies of printed material [30]. It is time consuming and expensive in terms of human resources to manually transcribe each document. In the case of historical manuscripts, this is more so due to the degradation of handwritten text [22] and sparsity of human resources, which adds more complexity to the task. Techniques that might work on a set of historical documents might not work on another. Thus, there is a large set of documents which are still not accessible because they yet remain undigitised. Through digitised documents, researchers that are working on old manuscripts can use machine translators to translate text and study the documents through semantic analysis techniques in order to extract knowledge for automatic text categorisation. In such a scenario, high-quality images of each page of the manuscript are taken to limit the damage that might affect the documents when handled for prolonged periods. For undigitised documents, the researchers initially transcribe the documents before applying any analysis techniques. For short manuscripts, this might not be an issue; however, large volumes of text would require a substantial amount of time and resources to transcribe properly. Furthermore, the said researchers might be looking for specific content within the text. Simple tasks such as keyword finding would be impossible unless the text is initially digitised [41]. On the other hand, with the use of an OCR or HCR system (depending on the type of text), the text could be digitised within a reasonable amount of time, and the researchers can then focus on the study of the text rather than the medial task of transcribing to a digital text document. In this chapter, we review the current approaches to HCR, starting with techniques that require segmenting documents into lines, words and characters as well as more modern techniques that do not require segmentation.

2 Handwritten Character Recognition

2.1 Offline Verses Online

HCR is divided into two sub-areas; the recognition of online text and offline handwritten text [39]. In online handwritten text analysis, the character recognition system deals with a constant stream of data which is produced by a transducer such as a sty-

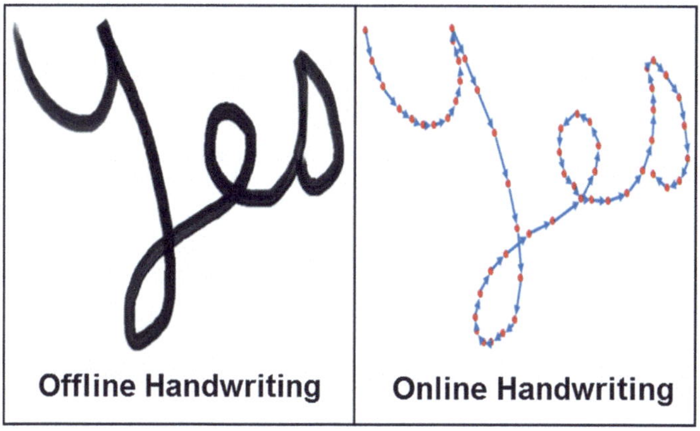

Fig. 1 (Left) Offline handwriting, (right) points produced by the stylus on online handwriting

lus pen. As the user is writing on a digitiser or tablet the device and the recognition software capture the recursive movements of the pen and acts upon these movements to capture the electronic signals emitted by the device as data that can be used to interpret the handwritten characters (see right) Fig. 1). In online handwriting recognition, recognition is done in real-time. This process facilitates the recognition somewhat as recursive loops and lines are easier to interpret as the user is writing [10]. On the other hand, offline character recognition is the process of interpreting and transcribing characters on handwritten text images. These images are usually pre-processed using several algorithms to facilitate segmentation and recognition. Offline text recognition is more difficult due to the state of the text data, and different authors have different handwriting styles [15]. Developing HCR systems that work on offline handwritten text usually requires text to be segmented into individual characters before the transcription can commence. This process affects the overall accuracy of the HCR system as errors introduced in segmentation might affect the final classification of the character as seen in (Fig. 2). By using machine learning techniques, specifically using deep learning techniques, such as recurrent neural networks, document images can be fed to a model as an input. The model then learns how to segment and transcribe characters. Thus, completely avoiding the process of segmentation. Although this approach has been used successfully, yielding positive results, several document images need to be manually transcribed in order to develop the initial training set as per [1]. In essence, this removes the process of automatically converting document images into machine-readable text as certain manual transcription needs to occur beforehand.

Fig. 2 Examples of incorrect character segmentation that might affect final character recognition

2.2 HCR Segmentation Techniques

Segmentation is a process that divides an image into its smallest parts. Each part is a component that is extracted from the previous block. In document segmentation, a document image is divided into lines, words and characters before finally passing through the character recognition process [4–6] (Fig. 3). This approach is also used in industrial document digitisation that makes use of these techniques for transcription and other document image analysis techniques such as keyword spotting. Document segmentation algorithms techniques are not limited to the segmentation of text lines, words and characters. Certain techniques are used to segment and extract contexts such as images, headlines and articles. In the following section, an overview of current techniques used in line, word and character segmentation are explored. The following section goes through some of the approaches used in segmentation to finally output a series of segmented characters needed for classification.

2.2.1 Pre-processing

Most of the challenges in handwritten character recognition are due to the composition of handwritten text. Different scribes and writers have their style of writing, which makes it difficult to create a system that can generalise for all these writing style variations. Moreover, the same manuscripts even when written by the same individual might still feature minute changes such as different text orientations, connected components in the text as well as document degradation. Pre-processing is used in order to try to normalise the documents before segmentation; thus, increasing the probability of successfully segmenting the documents. The approach used is also dependant on the segmentation techniques, a popular pre-processing technique con-

Fig. 3 Handwritten character recognition system

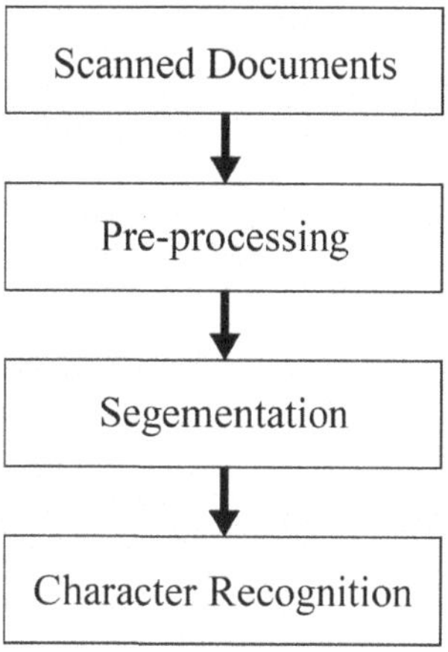

verts the image into black and white. This process called binarisation[1] helps separate the components of the document and augment the definition and curvature of the text (Fig. 3) [36, 46, 51] (Fig. 4). Text orientation is corrected by using techniques such as skew angle correction [47]. Other approaches include Hough Transform, introduced by Paul Hough in 1962 [16], and this technique uses parametric line representations to detect the lines within an image. The angle of the lines is then detected in order to calculate the mean angle to use when correcting a skewed document image [43]. Principal component analysis (PCA) is used in conjunction with binary images. The black pixels detected within an image are represented in a 2-dimensional vector. From this representation, a projection profile can be extracted to represent a set of 2-dimensional black pixel vectors. These vectors will point towards the direction of rotation of the image. Thus, detecting the skew angle difference [2] (Fig. 5). Median filters and denoising filters are used in order to correct document degradation. Sometimes the quality of the scanned images and the processing applied to images often create unwanted degradation. This degradation or noise can severely affect segmentation algorithms. Simple filters such as Gaussian blur and Median blur are widely used in the denoising process of document images. Applying a Gaussian blur to an image is simply the process of convolving an image with a 2-dimensional vector where the pixel values add up to a Gaussian distribution. This process introduces a smoothing effect that also softens edges and other details within an image [21]. Median filters, on the other hand, are applied by using a sliding window that goes through an image and replaces each pixel with the median value of the neighbouring pixels. The size of the sliding window can be changed depending on the amount of 'smoothing' needed to correct the image. 'Skeletonisation' is a technique applied to binary images that reduce the thickness of a stroke to one pixel. Zhang-Suen is the most popular application of the 'skeletonisation' technique. This method works by going through each pixel within a binary image. If a black pixel is found, the algo-

[1] Binarisation [38].

Fig. 4 a Unbinarised document images, b binarised document images [18]

rithm checks whether all eight neighbouring pixels are also black. If this condition is satisfied, then the neighbouring pixels are set to a white colour [50] as seen in Fig. 6.

2.2.2 Line, Word and Character Segmentation

Different components that makeup document images use different techniques. In this section, an overview of techniques used at each level of the segmentation process is explored. Some of the techniques are common across all segmentation levels. One such technique is the projection profile. These techniques use histograms to extract white space regions within a document image. These algorithms require that the document images are converted to a binary representation and correctly orientated.

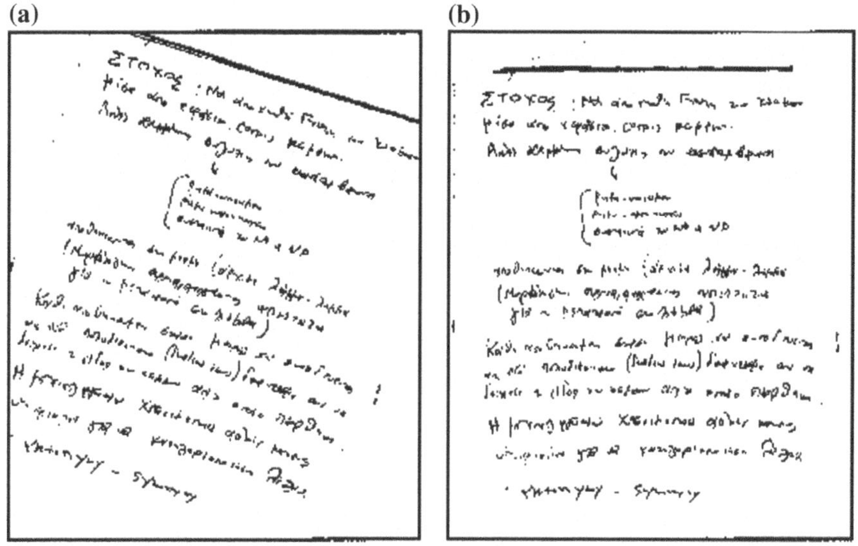

Fig. 5 **a** Incorrectly oriented text document, **b** correctly oriented text document

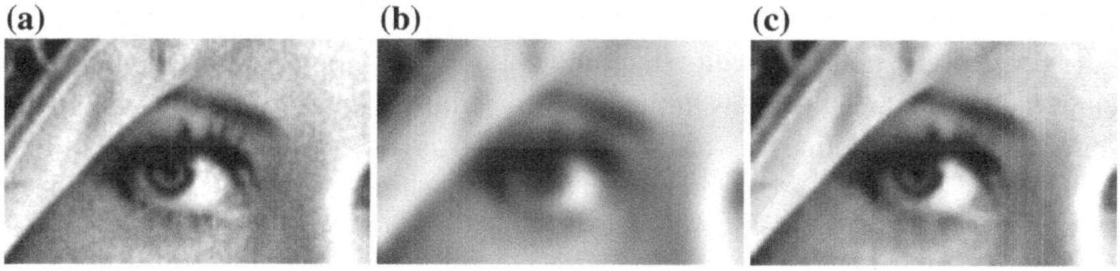

Fig. 6 **a** Noisy, **b** gaussian filter, **c** median filter

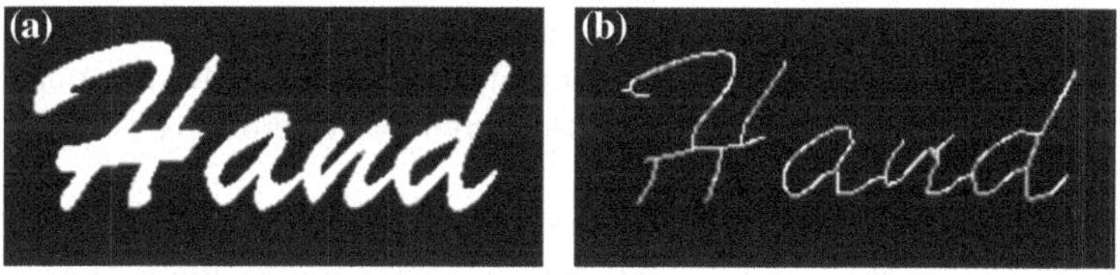

Fig. 7 **a** Original binary image, **b** skeleton of the image

The white lines dividing the text lines are used to segment the lines. The problem with using projection profiles is that in the case of handwritten text, it is not always the case that lines can be separated by a straight line [37]. Moreover, in handwritten text, text lines might overlap. Thus, no correct straight line can be computed that can completely segment lines, words and characters. The same process is also applied to word segmentation and character segmentation with various degrees of success [25, 28] (Fig. 8).

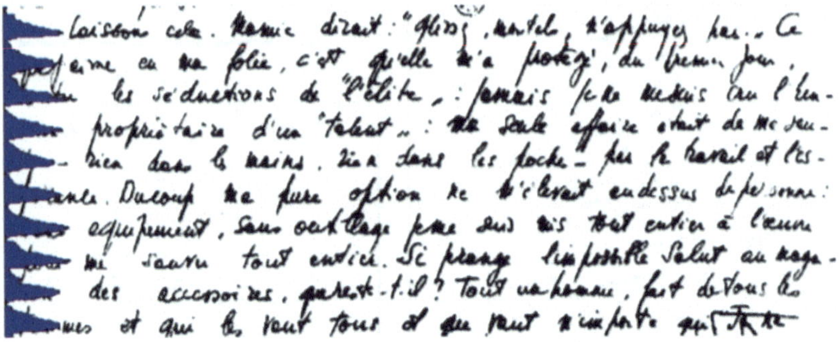

Fig. 8 Projection profile segmentation applied to an autograph of Jean-Paul Sartre

Unlike projection profile techniques, white space analysis methods calculate the space found between lines and words in a document image by computing a histogram measuring the occurrence of white pixels in a binarised image. White space analysis calculates minimum space found between connected components. These connected components are then grouped horizontally to create chains that represent the lines or vertically to represent words. This process leads to grouping the connected components vertically to compute white lines and column dividers [8]. This technique has achieved excellent results but is limited only to certain types of documents where the white space clearly defines a word separation such as in machine printed text or historical manuscripts that are written in an almost machine-like script. This approach has won various ICDAR competitions.

Algorithms using mathematical morphology make use of erosion and dilation techniques[2] to segment the lines within an image. This technique has proved to be quite successful, notwithstanding how straightforward and straightforward the approach is. The assumption when using mathematical morphology is that the document is divided into horizontal lines. Other techniques combine smoothing techniques, such as Gaussian blur to remove noise produced by erosion and dilation and improve line extraction [4, 5] (Fig. 9).

Filters are used mostly in Hybrid segmentation techniques. Nonetheless, the authors in [42] developed an algorithm called the Steerable Directional Local Profile Technique. This technique is based on observations they made on the human process of extracting lines. Humans tend to extract lines from a document image by detecting text line patterns. Thus, an adaptive local connectivity map (ALCM) feature is used to change the scale of a document image. Horizontal directions are computed for each

[2]Erosion and dilation techniques [45].

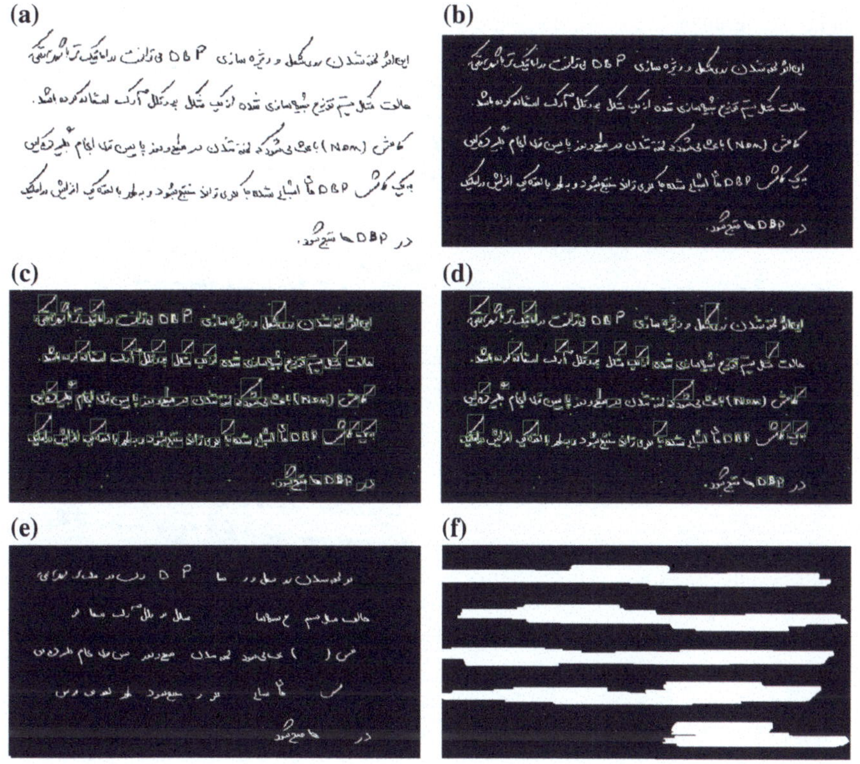

Fig. 9 Application of Erosion and Dilation techniques (**a**, **b**, **c**, **d**, **e**) followed by Gaussian blur (**f**) to extract lines

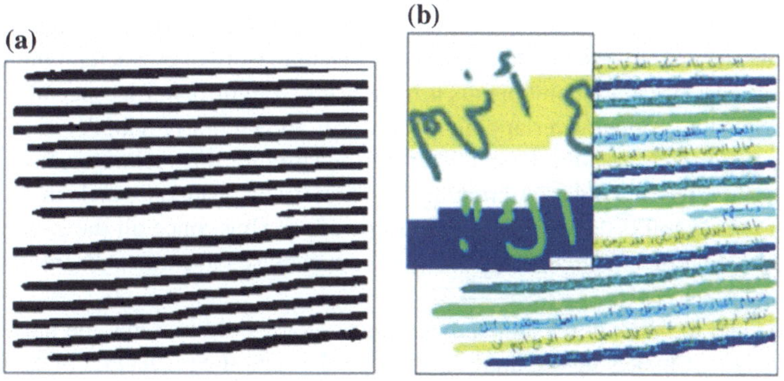

Fig. 10 ALCM connectivity map applied to handwritten text

pixel using the intensities measured by connecting neighbouring pixels. From the extracted ALCM, a local adaptive thresholding algorithm is used to identify the text line patterns found in the document image with respect to their connected component (Figs. 7 and 10).

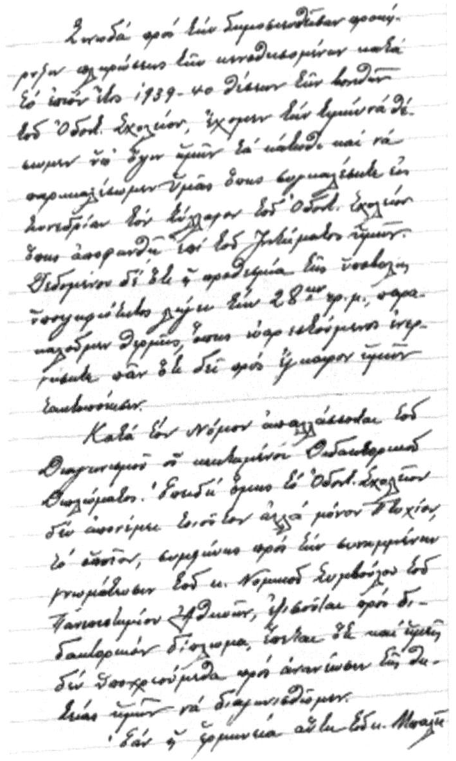

Fig. 11 The result of correctly applied hough transform technique applied to handwritten text [27]

Edge detection is a technique used in computer vision to identify the edge of an object or shape inside an image. These techniques have also been applied successfully to handwritten text in order to detect the edge of a line, word or character. Finding the edge that separates these components is not always straightforward, especially given manuscripts that contain overlapping lines, words and recursive text. A typical edge detection approach uses the Hough Transform. Through this technique, each edge pixel is converted to a polar co-ordinate. The premise is that edge pixels mapped on the same *theta* and distance r lie on the same edge. Thus, once all the edge pixels have been converted to polar form, the values are analysed to extract all straight lines within the image [24].

If one had to break down a document image, you would find that the most basic component that makes up the image are the strokes and the pixels that make up the text. Computationally finding all the strokes within an image can be very intensive. Clustering techniques measure the distance between connected components at the horizontal and vertical positions. The algorithm then assumes that short distances between connected components on the vertical space are on the same line. Similarly, distances of connected components on the horizontal line are assumed to make up the following or previous text line. Yin et al. [49] made an important observation that it

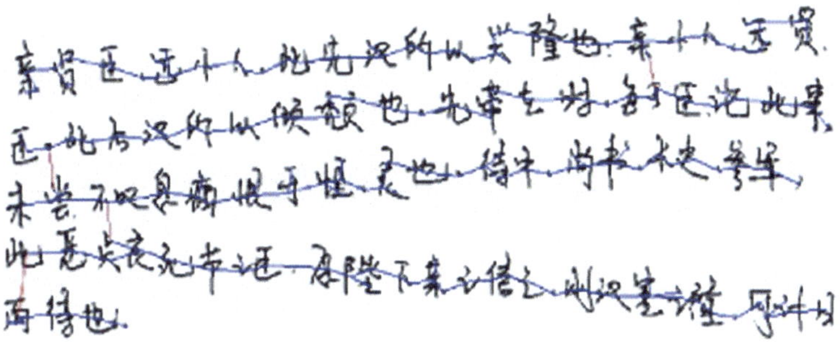

Fig. 12 Minimum Spanning Tree (MST) clustering algorithm applied to Chinese characters [49]

is not always the case that a line is oriented in the same direction. This is because text lines do not have a regular border defining their position. Thus, through the use of the features extracted from the connected components, certain heuristics are formed. For example, the distance between the connected component falling on the same line is smaller than the computed distance between alternate lines. This heuristic is not true for all cases as overlapping lines invalidate the claim. On the other hand, by using supervised learning algorithms, a learning component can be trained to distinguish connected components that are overlapping and segment accordingly. Each component is then aggregated in a component of a graph. Nodes within the graph are connected with an edge, where the distance between the edges defines the weight. The lines are then extracted from the graph by searching for the straightest line described in the graph [26] (Fig. 12).

Function analysis is top-down algorithms that try to approximate the objective value (segmented line or word) to a function. Functions such as energy mapping, contours and probabilistic layout estimation are computed. These algorithms usually work with boundary edges rather than boundary areas. Bukhari et al. use the parametric snake model that moves through the document image to compute the minimal energy function and extract active contours [6]. The approach works well on curled printed documents Bukhari et al. [7] have also adapted this technique to handwritten text segmentation (see Fig. 11). The proposed adaptation works on the ridges found in the handwritten text which are captured by applying a gradient vector flow to a document image [48] (Fig. 13). Machine learning algorithms applied to document image segmentation work on different document image type, including black and white, greyscale and colour images. These algorithms can discern differences within the text in order to distinguish features such as textual and generic features. Machine learning is a set of statistical models that through training, can learn how to perform a task.[3] In this case, the models learn how to segment lines, words and characters from document images. Diem et al. [11] use Gradient Shape Features (GSF) extracted from grey-level images to distinguish between handwritten and printed text. Classification is achieved by using a support vector machine (SVM) (Fig. 14). A GSF

[3]Machine learning [35].

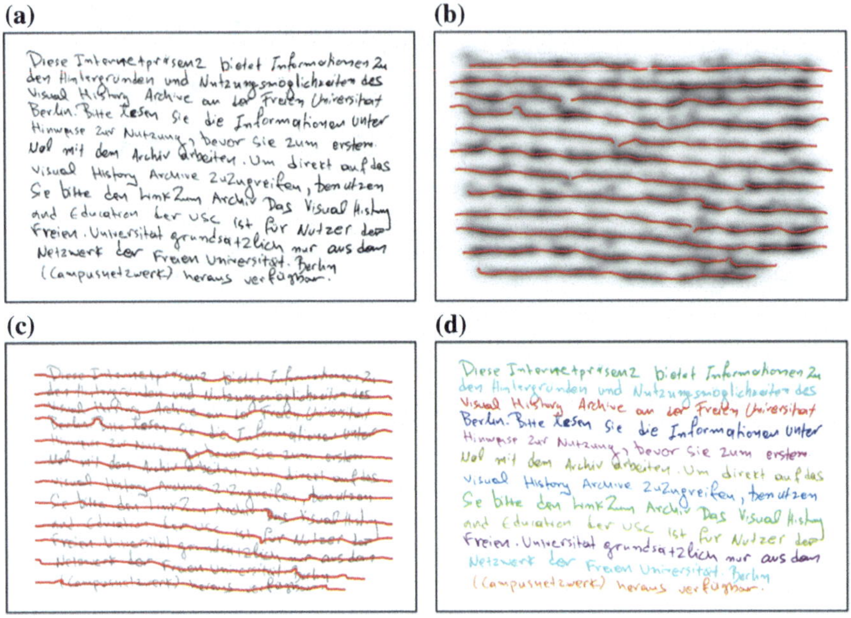

Fig. 13 Script-independent handwritten text line segmentation using active contours applied to ICDAR 2007 contest dataset [7]

matrix is built with the size of a typical word blob that goes through the image to extract word components. These components are then geometrically clustered to extract the lines. The final classification is achieved by using another SVM model that assigns the appropriate label to the final output. Other approaches include Huanfeng et al. [29] that also use SVM but train their models using Gabor filters and Gaussian mixture models applied on full-colour document images.

The final approach uses a subset of machine learning models. A neural network is a machine learning approach that groups different machine learning models to form a complex network that can learn how to perform a non-specified task through examples. The model itself manages to learn what it needs to do by comparing the input with respect to the corresponding output. Recurrent neural networks (RNN) are a class of feedback based neural networks combined with long short term memory (LTSM) activations to segment document images. Pre-labelled images are used to train to a neural network to identify the location of the first character for each line within the unseen dataset of document images. The predictions resulting from the neural net classify lines and interlines, and a post-processing step is applied to extract the final text lines [32, 33] (Fig. 15).

Fig. 14 A page from the PRImA database, green rectangles indicate correct segmentation, red rectangles indicate incorrect segmentation

2.2.3 Character Recognition

In this section, an overview of different AI classification algorithms is presented. These models assume, that, the input is a set of correctly segmented character images. In machine learning the feature set used for training the model is important in determining the overall classification accuracy. Depending on the number of features, the size of the dataset and the quality of the features, the model's accuracy is increased. Moreover, the larger and more varied the dataset, the more generalised the model becomes. Some of the features used in handwritten recognition include diagonal features, contour features and geometrical features. Diagonal features are extracted by dividing the image into bins and extracting the number of diagonals found at each bin (Fig. 16).

Counter features are extracted using counter detection algorithms such as Freeman Chain Codes (FCC). These algorithms extract the edge direction of a character image (Fig. 17). A normalisation procedure is then applied to normalise the number of features extracted. Dingli et al. [12] applied Elliptic Fourier Feature Normalization and improved the overall accuracy of the system with respect to standard count normalisation commonly applied to Free Man Chain Code feature set. The model used is a Deep Neural Network a class of Neural Networks that trains on large data sets. Nasien et al. [34] use Freeman Chain Code Scheme in their work on character recognition. The features also include a random based approach as a heuristic for normalisation coupled with secondary features including the ratio of width to height, ratio-upper with respect to the whole character, as well as ratio-right and ratio-

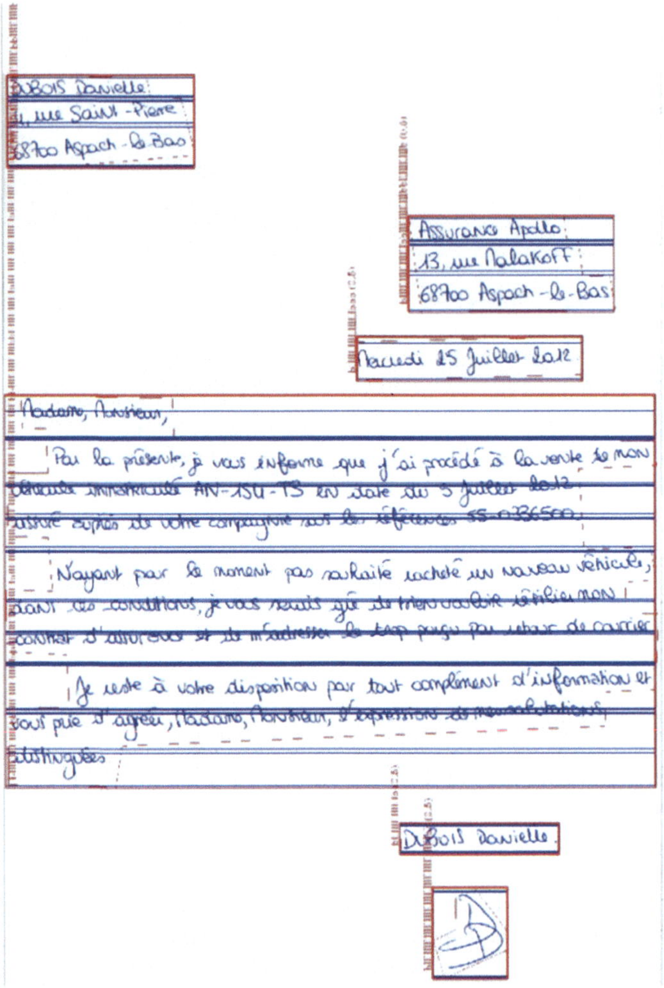

Fig. 15 RNN approach applied to a handwritten letter document image [32]

height. The dataset used is the NIST dataset. The authors also reported problems with the dataset given that some of the images were of poor quality and pre-processing steps such as thinning resulted with broken characters. Freeman code extraction was impossible to extract from the broken character images and was omitted from the final feature set. Furthermore, the samples were divided into sets of lowercase, uppercase and finally, a set of mixed lower and uppercase characters. An SVM classifier was used with a radial basis function set as the kernel. The final feature set included 169 features. The reported accuracy for the system was as follows 86% accuracy for lower case dataset, 88% accuracy for the uppercase and 73% for the combined dataset [34]. Gaur et al. [14] use SVM classifier for the work in handwritten Hindi character

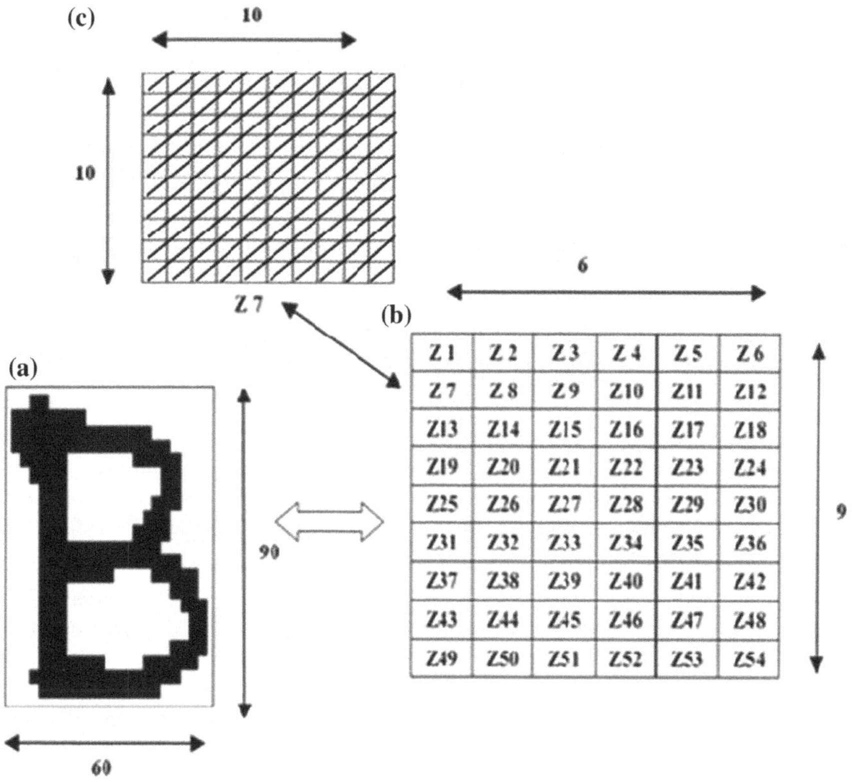

Fig. 16 **a** 10 × 10 bin with diagonals, **b** set of bins mapping the image, **c** original handwritten character image

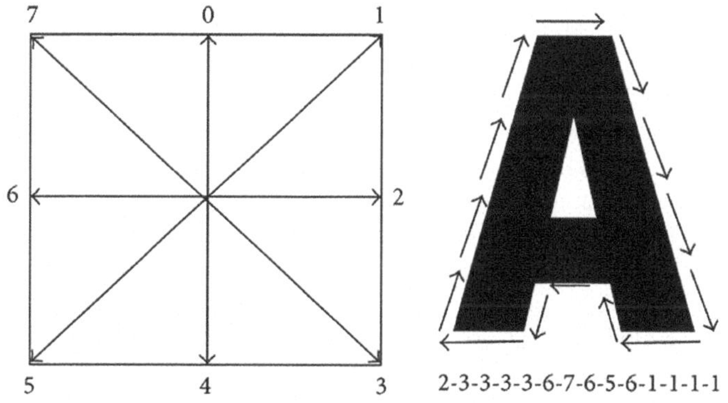

Fig. 17 An eight-directional FCC applied to letter ‘A’

recognition. K-means clustering is used on a binary image to extract features. The advantage of K-means clustering is that more robustness can be achieved for low illumination. Moreover, given that the dimensions of the feature set are relatively low computational overhead is reduced. For each binary image in the data set, K-means clustering is applied. The binary images were resized to 70×50 pixels. For each cluster, the pixel positions are extracted, and the pixel density is calculated. Thus, this will result in a single value representing each cluster. The cluster values are then ordered by their x and y co-ordinate value starting from the first cluster on the top left part of the image in order to create a feature vector of 35 values. For classification, SVM was used in combination with a linear kernel. Recognition of 95.86% was achieved on a dataset of 290 images with 140 images used for training and 150 images for testing. The SVM classifier was also used by Kadhm et al. [19] in his work on handwritten Arabic text recognition. In their work, the authors can classify characters using word images instead of a character image. A grey scaled image of the word is binarised using a Fuzzy C-Means clustering. A 3×3 median filter is used to denoising the image from the noise introduced by the binarisation process. The images are further pre-processed using thinning and resized to a 128×128 pixel image size. Three distinct features are extracted. Structural features are also extracted for each word image. Extracted features consist of zigzags, dots, loops, endpoints, intersection points and directional strokes. Statistical features are also extracted, which include the number of connected components and zoning features. The latter is extracted by dividing the word image into four separate zones were at the diagonals of each zone the summation of the black pixels is calculated and added as a vector of features. Finally, the Discrete Cosine Transform (DCT) features are extracted. This feature reduces the dimensionality of the image into a feature set of fifty coefficients. Normalisation is also applied to the whole feature vector to set to normalise to values to a range from 0 to 1. Linear, polynomial, sigmoid and RBF kernels are used. From a data set of 2913 handwritten Arabic word images, 70% of the images were used for training and 30% testing. The authors reported that the system achieved 96.3% accuracy when combined with the polynomial kernel.

Convolutional Neural Networks (CNN) are feedforward networks. These networks have achieved excellent results in areas of image recognition. CNN use the concept of convolution in order to create an architecture using several layers of convolution and nonlinear activation functions [20, 44]. Yann LeCun pioneered CNN in 1988 [23]. The resulting architecture called LeNet was used for character recognition for zip codes and numerical digits. CNN is attractive in the context of image recognition as they do not require feature extraction. This attractiveness is due to the ability of these models to directly process pixels and is thus trained using images. Unlike the previous models, the quality of the feature extraction methods used does not affect the final accuracy. The model thus creates its interpretation of the images and extracts features that it internally uses to discern different characters. Simard et al. [44] developed a CNN for handwritten digit recognition that uses an augmenting dataset. Dataset augmentation is the process of adding distortion to the dataset to improve the generalisation obtained by the neural net (Fig. 18).

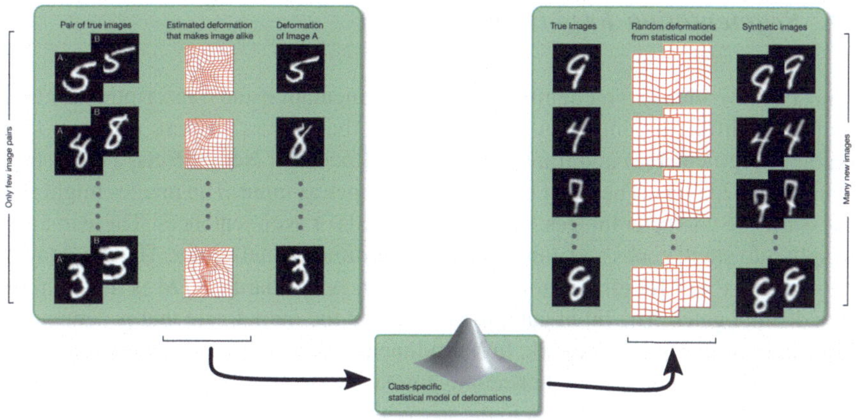

Fig. 18 Character dataset augmentation process

The MNIST dataset was used for training and testing. The dataset was split into 60,000 images for training and a further 10,000 for testing. By using the augmented dataset, they managed to obtain an overall error of 0.4% which was considered as state of the art at the time. An augmented dataset is also used in the CNN proposed by Ciresan et al. [9]. The architecture of the model uses an input layer of images size 29×29, where the original images are resized from 128×128 to 20×20 size image and centred over a 29×29 blank image. A convolutional layer with a 4×4 kernel followed by a max pooling layer with a kernel size of 2×2 connected to a 9×9 kernel convolution layer and connected to a final max pooling layer with a kernel size of 3×3. The final fully connected layer made up of 150 nodes and a final classification layer with a varying number of nodes depending on the dataset used—62 classes for the NIST SD dataset and 10 for the MNIST dataset. The dataset is distorted at the beginning of every epoch iteration using an elastic deformation with a variable value of $\alpha = 6, 36$ as well as a vertical and horizontal scaling of range from 0 to 15% and distorted by $\pm 15\%$. An error of 11.8% is reported when classifying uppercase and lowercase letters. The presented architecture can classify around 10,000 characters per second. The above implementations trained convolutional neural networks to classify English handwritten text. The method proposed by Rahman et al. [40] was evaluated on a dataset of handwritten Bangla characters. The dataset contains 20,000 handwritten characters with 400 character images representing each character of the Bangla alphabet. The input player of the network accepts a grey scale image with a dimension of 28×28 pixels. The architecture of the network used includes 2 convolutional layers using a 5×5 kernel, 2 max-pooling layers using a 2×2 kernel a fully connected layer with 192 nodes and the final output layer made up of 50 nodes representing all the letters in the Bangla alphabet. The dataset was split into 17500 character images for training and 2500 character images for testing. The system obtained an overall accuracy of 85.36% accuracy. The authors noted that although the results were promising, they did not compare to results achieved to other machine learning algorithms [40].

2.3 Segmentation Free HCR

In most handwritten text recognition systems, segmentation and transcription are two completely different components. In this paper, the authors propose a system that combines segmentation and transcription using Recurrent Neural Nets. The system is built up of 7 layers. The input layer accepts images as input. The images might not necessarily be character images, rather whole words or even sentences. These images are split into small zones and converted to a one-dimensional vector. The input layer is connected to a Multidimensional LSTM layer. A standard LSTM is made up of three distinct gates. The input gate, forge gate and the output gate that is connected to one Recurrent Neural Network. In Multidimensional LSTM, the connections have been extended to n recurrent connections for each of the nodes previous states. The resultant output is converted back to the size of the first zone and fed to a feedforward network which uses a tanh activation function. This process is repeated another two times up to which the final MDLSTM layer converts the output to a one-dimensional vector and transcribes to the Connectionist Temporal Classification layer. The latter output layer is specifically designed classification layer used in Recurrent Neural Nets where it transforms the output into sequence labelling and does not require pre-segmentation of labels or post-processing to convert to transcription values. The training data used were a set of 1518 images with 120 distinct characters. The system achieved an accuracy of 96.75% In his paper of the use of Recurrent Neural Networks as a handwritten text recognition system Jameel, made some observations on what input can be fed to the network [17]. Instead of feeding images, a set of features were extracted and used as input. Jameel argues that although curves, lines and intersections are intuitive features, to extract the sequence in which they appear is also very important. Thus, shadow features were extracted by computing a sequence of values that depict what happens when scanning a character image and at what time curves and other features appear in the image. The sequence was then used as a feature set to input into the recurrent neural network. A backpropagation neural network was used, that is, a fully connected recurrent neural network. The training set was made up of 877 character images, including uppercase and lower case letters. It took from 10,000,000 to 15,000,000 steps to train the network and achieve 91.4% accuracy [17].

3 Conclusion

In this chapter, an introductory overview of different techniques and approaches used in HCR are presented. An amalgamation of these different techniques will probably need to be used when trying to transcribe historical documents. The process might not be worth it at the moment when applied to a small dataset of a few pages. A well-constructed process might yield excellent results when applied to a set of documents that are either written by the same author or written using a particular writing style.

Also, most of the techniques can be used to affect simple keyword searches on the text document. If enough words are correctly transcribed then Natural Language Processing techniques can be used to extract keywords and semantics from the text without requiring full transcription.

References

1. I. Ahmad, G.A. Fink, Training an arabic handwriting recognizer without a handwritten training data set, in *2015 13th international conference on document analysis and recognition (ICDAR)* (IEEE, 2015), pp. 476–480
2. M. Basavanna, S. Gornale, Skew detection and skew correction in scanned document image using principal component analysis (2015)
3. D.C. Blair, M.E. Maron, An evaluation of retrieval effectiveness for a full-text document-retrieval system. Commun. ACM **28**(3), 289–299 (1985)
4. T.C. Bockholt, G.D. Cavalcanti, C.A. Mello, Document image retrieval with morphology-based segmentation and features combination, *Document Recognition and Retrieval XVIII*, vol. 7874 (International Society for Optics and Photonics, 2011), p. 787415
5. C.-A. Boiangiu, M.C. Tanase, R. Ioanitescu. Text line segmentation in handwritten documents based on dynamic weights. J. Inf. Syst. Oper. Manag. 1 (2013)
6. S. Bukhari, F. Shafait, T. Breuel, Segmentation of curled textlines using active contours (2008), pp. 270–277
7. S.S. Bukhari, F. Shafait, T.M. Breuel, Script-independent handwritten textlines segmentation using active contours, in *2009 10th International Conference on Document Analysis and Recognition* (2009), pp. 446–450
8. K. Chen, F. Yin, C.-L. Liu, Hybrid page segmentation with efficient whitespace rectangles extraction and grouping, in *2013 12th International Conference on Document Analysis and Recognition (ICDAR)* (IEEE, 2013), pp. 958–962
9. D.C. Ciresan, U. Meier, L.M. Gambardella, J. Schmidhuber, Convolutional neural network committees for handwritten character classification, in *2011 International Conference on Document Analysis and Recognition (ICDAR)* (IEEE, 2011), pp. 1135–1139
10. S.K.S. Dalbir et al., Review of online & offline character recognition. Int. J. Eng. Comput. Sci. **4**(05) (2015)
11. M. Diem, F. Kleber, R. Sablatnig, Text classification and document layout analysis of paper fragments, in *2011 International Conference on Document Analysis and Recognition (ICDAR)* (IEEE, 2011), pp. 854–858
12. A. Dingli, M. Bugeja, D. Seychell, S. Mercieca, Recognition of handwritten characters using google fonts and freeman chain codes, in *International Cross-Domain Conference for Machine Learning and Knowledge Extraction* (Springer, Berlin, 2018), pp. 65–78
13. L. Fortunati, J. O'Sullivan, Situating the social sustainability of print media in a world of digital alternatives. Telematics and Informatics (2018)
14. A. Gaur, S. Yadav, Handwritten hindi character recognition using k-means clustering and svm, in *2015 4th International Symposium on Emerging Trends and Technologies in Libraries and Information Services (ETTLIS)* (IEEE, 2015), pp. 65–70
15. A. Graves, J. Schmidhuber, Offline handwriting recognition with multidimensional recurrent neural networks, in *Advances in Neural Information Processing Systems* (2009), pp. 545–552
16. P.V. Hough, Method and means for recognizing complex patterns, Accessed 18 Dec 1962. US Patent 3,069,654
17. A. Jameel, Experiments with various recurrent neural network architectures for handwritten character recognition, in *1994 Proceedings of Sixth International Conference on Tools with Artificial Intelligence* (IEEE, 1994), pp. 548–554

18. P. Jana, S. Ghosh, S.K. Bera, R. Sarkar, Handwritten document image binarization: an adaptive k-means based approach, in *2017 IEEE Calcutta Conference (CALCON)* (IEEE, 2017), pp. 226–230
19. M.S. Kadhm, A.P.D.A.K. Abdul, Handwriting word recognition based on svm classifier. Int. J. Adv. Comput. Sci. Appl. **1**, 64–68 (2015)
20. A. Krizhevsky, I. Sutskever, G.E. Hinton, Imagenet classification with deep convolutional neural networks, in *Advances in Neural Information Processing Systems* (2012), pp. 1097–1105
21. B.S. Kumar, Image denoising based on non-local means filter and its method noise thresholding. Signal Image Video Process. **7**(6), 1211–1227 (2013)
22. V. Lavrenko, T.M. Rath, R. Manmatha, Holistic word recognition for handwritten historical documents, in *First International Workshop on Document Image Analysis for Libraries, 2004. Proceedings* (IEEE, 2004), pp. 278–287
23. Y. LeCun, B. Boser, J.S. Denker, D. Henderson, R.E. Howard, W. Hubbard, L.D. Jackel, Backpropagation applied to handwritten zip code recognition. Neural Comput. **1**(4), 541–551 (1989)
24. L. Likforman-Sulem, A. Hanimyan, C. Faure, A Hough based algorithm for extracting text lines in handwritten documents, in *Proceedings of 3rd International Conference on Document Analysis and Recognition*, vol. 2 (1995), pp. 774–777
25. L. Likforman-Sulem, A. Zahour, B. Taconet, Text line segmentation of historical documents: a survey. Int. J. Doc. Anal. Recognit. (IJDAR) **9**(2–4), 123–138 (2007)
26. L. Liu, Y. Lu, C.Y. Suen, Near-duplicate document image matching: A graphical perspective. Pattern Recognit. **47**(4), 1653–1663 (2014)
27. G. Louloudis, B. Gatos, C. Halatsis, Text line detection in unconstrained handwritten documents using a block-based Hough transform approach, in *Ninth International Conference on Document Analysis and Recognition (ICDAR 2007)*, vol. 2 (2007), pp. 599–603
28. G. Louloudis, B. Gatos, I. Pratikakis, C. Halatsis, Text line and word segmentation of handwritten documents. Pattern Recognit. **42**(12), 3169–3183 (2009)
29. H. Ma, D. Doermann, Word level script identification for scanned document images, *Document Recognition and Retrieval XI*, vol. 5296 (International Society for Optics and Photonics, 2003), pp. 124–136
30. R. Manmatha, N. Srimal, Scale space technique for word segmentation in handwritten documents, in *International Conference on Scale-Space Theories in Computer Vision* (Springer, Berlin, 1999), pp. 22–33
31. R.J. Mooney, L. Roy, Content-based book recommending using learning for text categorization, in *Proceedings of the Fifth ACM Conference on Digital Libraries* (ACM, 2000), pp. 195–204
32. B. Moysset, C. Kermorvant, C. Wolf, J. Louradour, Paragraph text segmentation into lines with recurrent neural networks, in *2015 13th International Conference on Document Analysis and Recognition (ICDAR)* (IEEE, 2015)
33. M. Murdock, S. Reid, B. Hamilton, J. Reese, Icdar 2015 competition on text line detection in historical documents, in *2015 13th International Conference on Document Analysis and Recognition (ICDAR)* (IEEE, 2015), pp. 1171–1175
34. D. Nasien, H. Haron, S.S. Yuhaniz, Support vector machine (svm) for english handwritten character recognition, in *2010 Second International Conference on Computer Engineering and Applications (ICCEA)*, vol. 1 (IEEE, 2010), pp. 249–252
35. N.M. Nasrabadi, Pattern recognition and machine learning. J. Electron. Imaging **16**(4), 049901 (2007)
36. N. Otsu, A threshold selection method from gray-level histograms. IEEE Trans. Syst. Man Cybern. **9**(1), 62–66 (1979)
37. N. Ouwayed, A. Belaïd, A general approach for multi-oriented text line extraction of handwritten documents. Int. J. Doc. Anal. Recognit. (IJDAR) **15**(4), 297–314 (2012)
38. L.O. Gorman, Binarization and multi thresholding of document images using connectivity. CVGIP: Graph. Models Image Process. **56**(6), 494–506 (1994)

39. R. Plamondon, S.N. Srihari, Online and off-line handwriting recognition: a comprehensive survey. IEEE Trans. Pattern Anal. Mach. Intell. **22**(1), 63–84 (2000)
40. M.M. Rahman, M. Akhand, S. Islam, P.C. Shill, M.H. Rahman et al., Bangla handwritten character recognition using convolutional neural network. Int. J. Image, Graphics Signal Process. (IJIGSP) **7**(8), 42–49 (2015)
41. T.M. Rath, R. Manmatha, Word spotting for historical documents. Int. J. Doc. Anal. Recognit. (IJDAR) **9**(2–4), 139–152 (2007)
42. Z. Shi, S. Setlur, V. Govindaraju, A steerable directional local profile technique for extraction of handwritten arabic text lines, in *10th International Conference on Document Analysis and Recognition, 2009. ICDAR'09* (IEEE, 2009), pp. 176–180
43. B.K. Shukla, G. Kumar, A. Kumar, An approach for skew detection using hough transform. Int. J. Comput. Appl. **136**(9), 20–23 (2016)
44. P.Y. Simard, D. Steinkraus, J.C. Platt, Best practices for convolutional neural networks applied to visual document analysis, in *null* (IEEE, 2003), p. 958
45. P. Soille, Erosion and dilation, *Morphological Image Analysis* (Springer, Berlin, 2004), pp. 63–103
46. M.H.J. Vala, A. Baxi, A review on otsu image segmentation algorithm. Int. J. Adv. Res. Comput. Eng. Technol. (IJARCET) **2**(2), PP–387 (2013)
47. A. Al-Khatatneh, S.A. Pitchay, M. Al-Qudah, A review of skew detection techniques for document, in *2015 17th UKSim-AMSS International Conference on Modelling and Simulation (UKSim)* (IEEE, 2015), pp. 316–321
48. C. Xu, J.L. Prince, Snakes, shapes, and gradient vector flow. IEEE Trans. Image Process. **7**(3), 359–369 (1998)
49. F. Yin, C.-L. Liu, Handwritten Chinese text line segmentation by clustering with distance metric learning. Pattern Recognit. **42**(12), 3146 – 3157 (2009). New Frontiers in Handwriting Recognition
50. T. Zhang, C.Y. Suen, A fast parallel algorithm for thinning digital patterns. Commun. ACM **27**(3), 236–239 (1984)
51. M. Ziabari, V. Mottaghitalab, A. Haghi, Application of direct tracking method for measuring electrospun nanofiber diameter. Braz. J. Chem. Eng. **26**(1), 53–62 (2009)

Supporting Autonomous Navigation of Visually Impaired People for Experiencing Cultural Heritage

Daniele Croce, Giovanni Galioto, Natale Galioto, Domenico Garlisi, Laura Giarré, Federica Inderst, Federica Pascucci and Ilenia Tinnirello

Abstract In this chapter, we present a system for indoor and outdoor localization and navigation to allow the low vision users in experiencing cultural heritage in autonomy. The system is based on the joint utilization of dead-reckoning and computer vision techniques on a smartphone-centric tracking system. The system is explicitly designed for visually impaired people, but it can be easily generalized to other users, and it is built under the assumption that special reference signals, such as colored tapes, painted lines, or tactile paving, are deployed in the environment for guiding visually impaired users along pre-defined paths. Differently from previous works on localization, which are focused only on the utilization of inertial sensors integrated into the smartphones, we exploit the smartphone camera as an additional sensor that, on one side, can help the visually impaired user to identify the paths and, on the other side, can provide direction estimates to the tracking system. The users with the help of the navigation system may experience the museum or the cultural site in auton-

---

D. Croce · G. Galioto · N. Galioto · I. Tinnirello (✉)
DEIM, Università degli Studi di Palermo, Viale delle Scienze, 9, 90128 Palermo, Italy
e-mail: ilenia.tinnirello@unipa.it

D. Croce
e-mail: daniele.croce@unipa.it

G. Galioto
e-mail: giovanni.galioto@unipa.it

N. Galioto
e-mail: natale.galioto@unipa.it

D. Garlisi
CNIT Consortium, Viale G.P. Usberti, 181/A, 43124 Parma, Italy
e-mail: domenico.garlisi@cnit.it

L. Giarré
DIEF, Università degli Studi di Modena e Reggio Emilia, Via P. Vivarelli, 10, 41125 Modena, Italy
e-mail: laura.giarre@unimore.it

F. Inderst · F. Pascucci
DING, Università degli Studi Roma Tre, Via della Vasca Navale, 79, 00146 Rome, Italy
e-mail: federica.pascucci@uniroma3.it

D. Seychell and A. Dingli (eds.), *Rediscovering Heritage Through Technology*, Studies in Computational Intelligence 859,
https://doi.org/10.1007/978-3-030-36107-5_2

omy, by following a path previously decided and by going from a location of interest to another, without any external personal assistant. We demonstrate the effectiveness of our approach, by means of experimental tests performed in a controlled indoor environment and in a real outdoor installation.

1 Introduction

In the last years we have assisted to many initiatives and experiments for strengthening the social role of museums, from a place mostly dedicated to the storage and exposition of art crafts, to a space of social dialogue and cultural activities, where museum interpretations and representations are often co-created by visitors and museum staff. In such a context, participation of vulnerable groups can be one of the key aspects to promote and reinforce their social inclusion. In this chapter we focus on people affected by visual problems. Visual impairment affects approximately 285 million people worldwide, of which 39 million are blind, and 246 million have moderate to severe visual impairment [1]. Estimates suggest that another person in the world goes blind every 5 s [2]. This is a large number of people who rely on a combination of their other senses—hearing, touch, and even smell—and tools like walking sticks and helpers. Improving museum accessibility for these groups of people is very relevant, especially taking into account that the impact of these disabilities is increasing with the aging of the world population.

Dealing with museum accessibility requires to face two different problems: extending the accessibility of the art experience, by finding innovative ways of perception beyond the sense of the sight; extending the physical accessibility of the spaces in which the art collections are placed. Most of the current research and experimentation efforts are focused on the first aspect. Indeed, people with visual impairment are excluded from directly experiencing the cultural heritage presented by museums in a traditional manner. A common approach for mitigating this problem is participating to the museum visits through audio descriptions and explanations of pieces, whether they are live (or recorded) commentaries provided by museums or provided from friends in real time. However, descriptions provided by others prevent one of the key aspect of art fruition, that is interpreting pieces of art by the users themselves. Innovative ways of perception are considering the exploitation of multimodal descriptions relying on multiple senses. Braille extensions and 3D reconstructions of paintings or building models, for example, have been proposed in many museums and exhibitions. Multimodal interaction allows visually impaired people to access cultural heritage involving large spatial information content, as described in [3] or [4], where the system makes it possible to interact with haptic/acoustic active objects and to select the information that must be shown on the basis of user requirements. But we really think this is only an aspect of the problem. As stated in a recent article appeared online [5]: *people are disabled more by inconsiderate design, inaccessible services, and other people's unaccommodating attitudes than by their own impairment.* Social inclusion of such large portion of population resides in letting them enjoy culture by a personal and autonomous experience, i.e. by allowing a low vision or blind person to enter in a museum, walk, visit and stop in front an artwork without a personal assistant.

The focus of the present chapter is concerning navigation systems for blind users, devised to encourage autonomous visits and improve physical accessibility of muse-

Fig. 1 An exemplary installation of the ARIANNA system

ums. Although indoor navigation systems are of wide interest for many different applications, it is worth noticing that solutions for blind people have stricter requirements than general systems, in terms of accuracy and reliability. For example, visually impaired people are used to a continuous reference signal, such as the one provided by tactile pavings, which guides the users along the path to the destination. This means that the navigation system cannot provide simple information on the directions towards a destination and new interfaces have to be considered. Our solution, called ARIANNA (pAth Recognition for Indoor Assisted NavigatioN with Augmented perception) [6–9], proposes to use the smartphone as a mediation instrument between the users and the reality. The main idea of the system is represented in Fig. 1: the smartphone camera detects the tapes in the environment using computer vision and provides user feedback in terms of vibration signals for identifying the paths. The system has been tested in permanent and temporary exhibitions in various cultural sites, such as the Farm Cultural Park (Favara, Italy), the GAM (modern art gallery) and the Branciforte Palace in Palermo (Italy), Expo 2015 in Milan, as well various other scientific events. The system has been also installed in the Blind institute of Palermo and tested by many low vision users. In the rest of the chapter, after a brief description of other possible solutions, we describe the technical features, localization performance and user experience results of ARIANNA.

2 Navigation Technologies for Blind People

Due to the pervasive availability of ICT technologies for localization, in recent years there was a proliferation of interesting solutions devised to improve the independence and social inclusion of visually impaired people [10]. In most cases, these solutions are focused on the design of specific user interfaces, rather than innovative localization solutions taking into account requirements for blind people. For example, Wayfinder offers an audio interface to a navigation system based on traditional GPS, providing directions and general descriptions about the outdoor places mapped by the application. Other systems working in indoor, where GPS cannot be used, are also based on general-purpose technological solutions, which that can be generally categorized into three approaches: triangulation of RF signals (mainly WiFi), direct sensing of reference points (implemented with RFIDs, ultrasound, bluetooth, etc.), ego-motion estimate provided by dead reckoning algorithms using *Inertial Measurement Units* (IMU, i.e., accelerometers, magnetometers, and gyroscopes). Examples of systems following these approaches and designed for blind people are: e-White, that uses WiFi or Bluetooth signals to coarsely locate the user's position and provide information on the nearest point of interest; RF-PATH-ID and Sesamonet, based on the detection of reference sensors whose presence is signaled by means of alert messages; Navatar [11], based on dead reckoning, where the user can provide references (such as a door in proximity) to improve the estimate of the position in a known map.

All these solutions have well known accuracy and practical limitations, which may represent a significant barrier for their adoption in museum itineraries. On one side, distance estimates by means of WiFi signals, which are available in most indoor environments for offering wireless connectivity, are affected by intrinsic positioning errors due to multipath; on the other side, reference points such as RFID or iBeacons may suffer of collisions in case of dense deployments and require a rigid environment instrumentation [12], not suitable for temporary exhibitions and frequent itinerary updates. In some cases, reference points can be represented by WiFi radio maps (also called fingerprinting), which need long and periodic calibration phases [13, 14]. Pedestrian Dead Reckoning (PDR) solutions for tracking user positions between consecutive reference points suffer from drift due to noise [15] on long itineraries, which can be only partially mitigated by filtering techniques [16], exploiting activity recognition [17], or periodic resets at the detection of reference points [12]. In other words, no system is currently able to provide a reliable and continuous reference signal along a museum itinerary, which can be considered equivalent to the tactical pavings.

Assistive technologies may also include other types of services for blind people, such as scene analysis and detection of obstacles. Obviously, these services are completely different from providing orientation signals towards a desired destination. Examples of systems working on obstacle detection are SmartCane [18] and UltraCane/Batcane [19], which integrate sonars and cameras. Other systems, such as the ones described in [20], offer robot-assisted navigation, in which decision-making is not left to the user, but it is rather performed by an artificial intelligence agent. Researchers have also evaluated novel interfaces for enabling such a guidance of a human user. For example, in [21] it is proposed the use of a vibration belt with distinct vibration patterns to communicate directional and rotational commands for navigation.

Our contribution is focused on the guidance service towards a desired destination, without any additional service on obstacle detection. We also propose an innovative interface for providing orientation information and a continuous reference signal along the paths. However, decision making is left to blind users, which are free to plan their visit in the museum itineraries according to their needs and experience.

3 The ARIANNA Navigation System

The ARIANNA navigation system allows autonomous mobility of blind people in public spaces, adopting a solution that is based on ICT (robotic and vision) technologies to mediate between users and environment. The system is especially designed for indoor scenarios, where GPS-based solutions are unavailable, and exploits the visual sensor and vibration signals of commodity smartphones. The system permits to find some points of interests in an indoor environment by following a path painted or sticked on the floor. The path is detected by the camera of the smartphone which also generates a vibration signal providing a feedback to the user for correct-

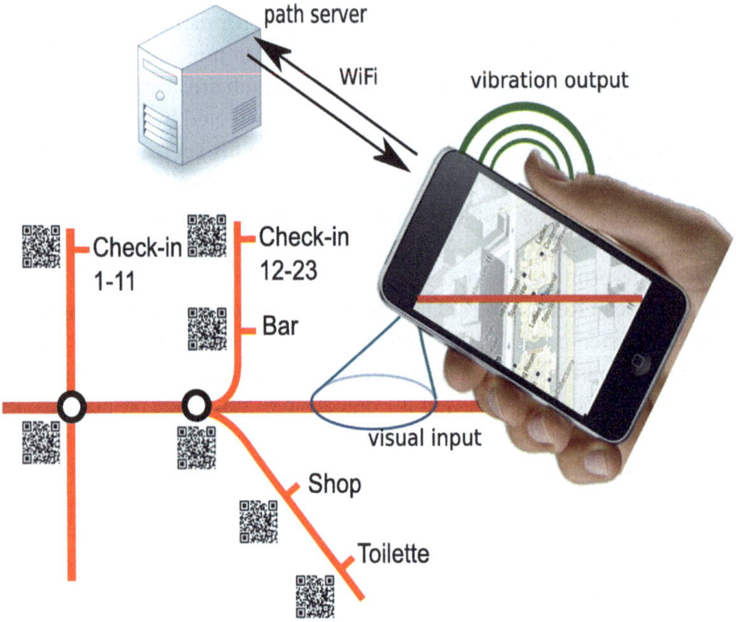

Fig. 2 ARIANNA navigation system description

ing his/her direction. Some special landmarks (e.g. QRcodes or iBeacons) can be deployed along the path for coding additional information detectable by the camera. All the computations and the computer vision algorithms run on the smartphone in real time.

Figure 2 highlights the different components of the system in an airport scenario. The paths of interest are marked with colored lines on the floor. This is the only dedicated instrumentation applied to the environment and is a quite simple and low cost solution; QRcodes are settled close to points of interest, on line intersections and are also used for landmarking. They provide information on the right line to follow in order to get to the desired destination. The user interface employs tactile stimuli to receive feedback on the heading corrections to be employed, as better described in the following. The systems itself is composed by five main components: (A) ambient instrumentation; (B) sensors; (C) data transport network; (D) path server; (E) tactile interface.

Ambient instrumentation. The ambient instrumentation is composed of colored tapes which can be easily sticked on the floor or carpets to define different paths. This is the only dedicated instrumentation applied to the environment. Paths can intersect each other forming a planar graph where intersections are nodes of the graph. To add information on the paths, any segment (the graph edges) may be deployed with two parallel strips with different colors, so the ordered couples (color1, color2) and (color2, color1) encode both direction and orientation. Additionally, using bar codes or QRcodes it is possible to encode relevant information regarding the edges (as for example the distance from/to the extremes of the segment) and for landmarking. Another possible solution is to deploy iBeacons to provide proximity information close to intersections or points of interests. QRcodes are generally more precise in terms of positioning, although aesthetically more visible.

Sensors. The main sensor used in the ARIANNA system is the camera, which most smartphones on the marketplace are equipped of. The camera is used to reveal the presence of lanes on the floor and acts as a visual control for the haptic transducer. We also use the embedded compass and accelerometer sensors to help maintain or recover the visual control of the line. All these sensors are available on most commodity smartphones: this is a key aspect for keeping the system low-cost and affordable for a vast public.

Data transport network. We assume that a data network connection is available for downloading the ambient map (e.g. via a WiFi or cellular network). The data transport network does not require specific adaptations but is a facility that permits communication between the phone and the ARIANNA server. The server is used to provide localization information, correlation between paths and points of interest, routing towards the destination. The presence of the server and the wireless network is necessary only in case the application is unaware about the building topology and its deployed paths. On the contrary, if the application loaded on the phone has such information locally available, the presence of network and server is optional (even if flexibility is possible only with those elements, as explained below).

The path server. The path server stores and retrieves information from a path repository via the url printed into the QRcode or provided by the iBeacon. The content pointed out by the (fixed) url can be changed on the fly with a simple update on the server. Such flexibility permits path adaptation required by topological changes due to maintenance or load balancing. When the smartphone detects a QRcode/iBeacon on the path, it immediately runs an http request to the server using the detected url. The server knows the position of the user (because of its proximity to the tag position) and sends back to the smartphone the next edge to follow. In fact, among all paths deployed in the building, thanks to the indications provided by the path server, the smartphones provides haptic feedback only towards the "enabled" paths according to the server indication.

Tactile Interface. The tactile interface is a key point of the system. The behavior of the haptic feedback can be summarized as follows: the camera continuously grabs the scene in front of the person. The tracking system incorporates the information on the line (together with the compass and accelerometer data) and provides feedback with the phone vibration. The intensity and type of the vibration is based on the output of the EKF and is designed to keep the camera always in contact with the line or to bring back the visual contact when it is lost. Vibration is a native functionality of the phone obtained through a rotating eccentric mass. It has been shown that the current consumption of typical vibration motors has a limited impact on the battery life of commercial smartphones [22] and that the energy savings coming from switching off the screen are higher than the costs introduced by vibrational cues [23]. Unlike other approaches in haptic interfaces, our solution does not need a selective vibration of the touched point (that is also difficult to obtain and requires special piezo-electric materials, etc.).

4 Computer Vision Algorithms

The main idea of ARIANNA is exploiting computing vision and vibration signals for replacing the special tiles deployed along the tactile pavings with easy to deploy colored tapes. There are many different computer vision functions that can be combined for the identification of a painted line, taking into account the constraints of our problem: (i) the path identification has to be prompt and reliable, without perceivable latencies for the users, which could correspond to discontinuous signals; (ii) the lifetime of the smartphone battery has to be compatible with the timing required for a complete museum visit, in order to guarantee the practical usage of the system. These constraints correspond to the identification of robust solutions, with limited complexity, able to work in real-time.

There are two main features that can be exploited for detecting the paths: the geometry of the tapes (which in the end are given by piecewise lines), and the colors of the tapes (which combine two different colors for representing a direction without ambiguity). We developed two different solutions, focused on both the geometric factors and the color identification, with different complexity and additional information that can be exploited by ARIANNA, as detailed in the following.

4.1 Geometry-Based Path Identification

An obvious solution for detecting a path is searching lines into the images, by using the well-known Canny algorithm, which is able to identify a set of edge points in an image delimiting areas with large luminance changes. This set of points can be associated to a path whenever they result aligned. Moreover, this kind of analysis allows to detect the slope of the identified path, which can be mapped into an heading measurement of the user movements along the path. In case a map of the paths is known and the user can be positioned (even roughly) on this map, the relative heading of the user can be converted into an absolute orientation.

To identify the line seen by the camera, we implemented three different steps: (i) filtering the image, for reducing the noise and the details of the image background; (ii) applying the Canny algorithm, for detecting the edges of the objects in the image; (iii) identifying the sub-set of edges which can be considered as a line using the Hough transform.

Eliminating image details. The first step is performed by using a Gaussian smoothing filter, whose main goal is defocusing the image for avoiding that some regular patterns of the floor (e.g., such as the edges of squared tiles) can be erroneously considered as a path trace. Since the lines deployed on the floor are very thick in comparison with the tiles' edges, such a filtering operation does not affect the identification of the line edges. The filter is characterized by a parameter σ which represents the standard deviation of the Gaussian function used for smoothing the image details. Higher values of σ lead to a more evident loss of image details.

Detecting edges. The second step is given by the application of the well-known Canny scheme. The output is a binary matrix, whose dimension is equal to the original frame and whose values are set to 1 for the pixels corresponding to the detected edges. These pixels are identified by computing the maximum luminosity gradient (in each possible direction) for each pixel, and by selecting the pixels for which the gradient is higher than a threshold T. Higher values of T correspond to a lower number of detected edges.

Detecting lines and slopes. The last step works on the binary image found by the Canny scheme for transforming the line identification problem in a maximum search problem. The Hough transform is used for mapping each edge point in a set of parameters (ρ, θ) representing the bundle of lines passing through that point. When multiple edge points are aligned, there is a common (ρ, θ) value representing the line passing through all the points. Therefore, the scheme simply works by counting the maximum number of occurrences (i.e., votes) of quantized (ρ, θ) values computed for each edge point. The quantization allows compensating noise effects in the identification of the alignments. This step is critically affected by the resolution used for quantization and by the threshold used for assuming that a given (ρ, θ) value corresponds to a line in the image. We chose to only consider the global maximum in the parameter space, i.e., a single (ρ, θ) value receiving the greatest number of votes, because we assume that the greatest number of aligned points always correspond to the thickly painted line. However, since the painted line is actually implemented with a bi-colored stripe, such a choice implies that we are only able to detect one of three parallel linear edges along the stripe. In proximity of a turning point, such as L-like junction points, the number of path segments with a different orientation is at least equal to two. However, by always choosing the line with the highest number of votes, which usually correspond to the longest detected line, we avoid ambiguity in the reference orientation for the user.

The above scheme can be optimized by considering the effects of the image resolution, the smoothing factor of the Gaussian filter, and the threshold values of the Canny schemes on the accuracy of line detection. Figure 3 visualizes the effects of different smoothing factors on a real image of a path trace. The standard deviation is normalized as a function of the image resolution (for representing the portion of the area involved in the filtering operations). As the standard deviation increases, the image loses details as evident in the reduction of the number of edges identified by the Canny algorithm. This, in turns, corresponds to a lower number of detected lines (from 9 lines for $\sigma = 0.25$ to 3 lines for $\sigma \in [0.5, 0.625]$), which are likely to coincide with the path trace, rather than with the floor regular patterns. Image resolutions have a similar effect on line detection accuracy: too high resolutions (apart from complexity issues) include too many details, which can correspond to the detection of wrong lines; too low resolutions imply rough estimates of the line orientation. We also noticed that the threshold of the Canny scheme has a limited

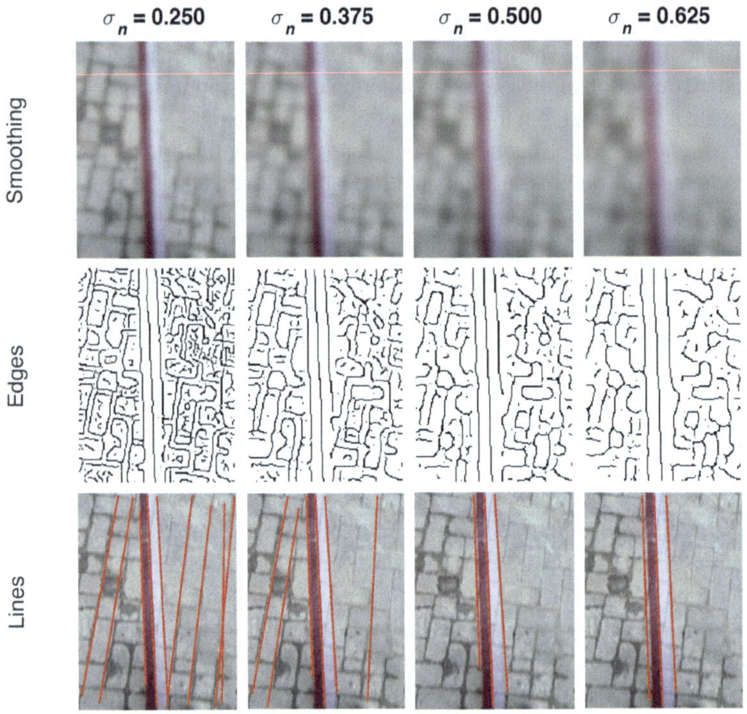

Fig. 3 Effects of the Gaussian smoothing filter on the capability of detecting the correct path trace

impact on the error metric. In the next experiments discussed in the paper, we used the optimal configurations found on the basis of these considerations.

4.2 Color-Based Path Identification

Another possible solution for identifying the path is to detect areas with the expected tape colors into the image. Image colors can have different representations, among which the HSV (hue, saturation, value) representation is more robust to luminance changes. In principle this approach can be quite simple, because it works by applying conditional statements on each image pixel, in order to verify if it belongs to the desired intervals of hue, saturation and values. The complexity required by this filtering operation is much lower than the complexity required by the previous approach, which requires many operations on the image matrices. Moreover, the direction information can be easily extracted from the blob of points identified as belonging to the path, by evaluating the second-order moments of the blob.

Despite of these considerations, the approach based on color searches has the drawback of being potentially affected by the light conditions in which the camera

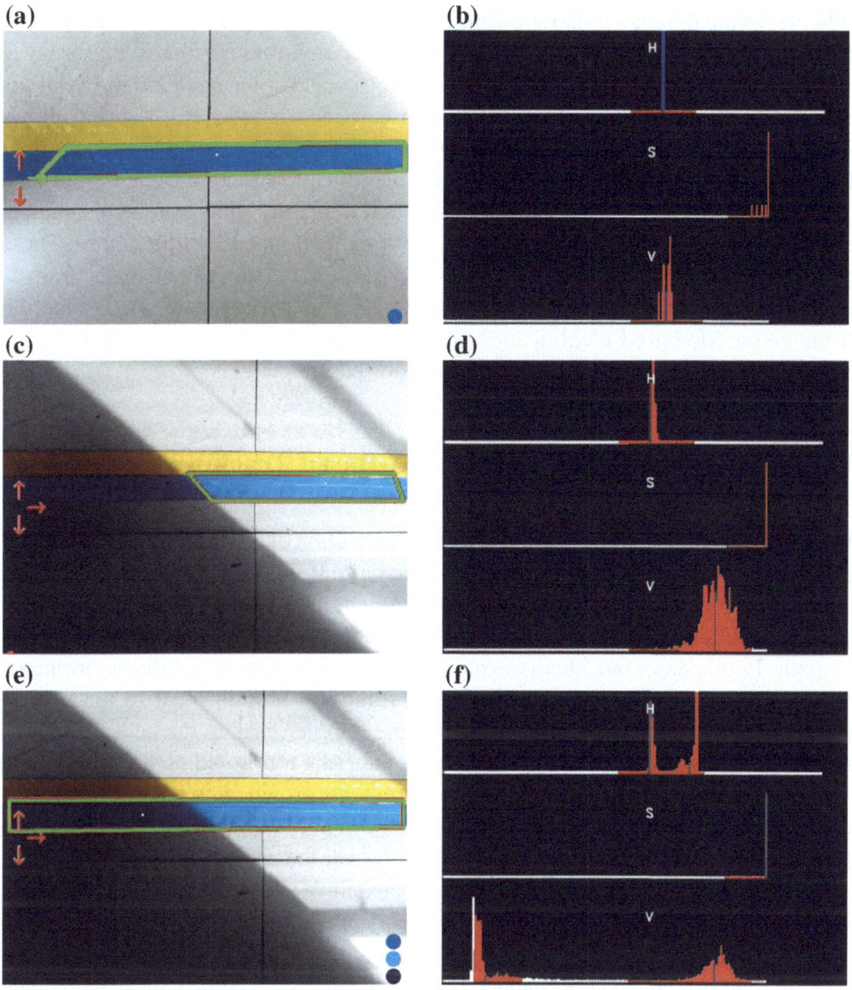

Fig. 4 Effects of the dynamic resampling of HSV filtering values

works. Indeed, it is not obvious how to define absolute HSV intervals for identifying the path under variable (even on a day-basis) light conditions. Schemes removing shadows on the images or correcting luminance changes can be more complex than edge-based solutions. We therefore implemented a mixed algorithm, in which the two schemes are opportunistically combined: ARIANNA normally works by applying a simple color-based detection, but at regular time intervals the geometry-based scheme is executed for delimiting the tapes and re-sampling the color intervals.

Figure 4 shows an example of dynamic resampling of the HSV values used by the color-based path identification approach. In the left part of the figure, we can see the bi-colored tape under three different light conditions. A red delimiter is used for segmenting the image area recognized as belonging to one color of the path. In the right part of the figure, we also show the HSV distributions of the points belonging to the segmented area. In the top case, when the light conditions are almost uniform, we can see that the HSV distributions are very narrow and color recognition by means of HSV filtering works well. In the middle case, we see that the HSV filtering allows to recognize only one part of the path, not covered by the shadowed area. Only after the execution of the geometric-based path identification function, the whole path captured in the image is correctly identified and the HSV values can be resampled, leading to the bi-modal distributions shown in the bottom case of the figure. These distributions can be used for applying the color filters to the subsequent image frames, until the user moves completely to the shadowed area. At this point, the next execution of the geometric-based identification function will update the HSV values to new single-modal distributions typical of the shadowed image.

5 Extending ARIANNA with IMU-Based Tracking

A possible question that could arise from the description of ARIANNA is understanding what happens whenever the user loses the path. If the camera is no more able to capture the colored tape, in some cases it is enough to sweep the smartphone around for finding it again. However, we cannot exclude that users accidentally abandon the path. In this case, our idea is exploiting the inertial sensors of the smartphone, i.e. the IMU systems, for supporting PDR solutions, able to localize the user in the space and provide information for bringing him/her back to the path. Basically, the IMU systems are utilized at regular time intervals for providing a direction and velocity measurement, by reading the measurements provided by the gyroscope and accelerometers of the smartphone. It is also important to estimate the user activity, in order to stop the measurement acquisition whenever the user stops walking. Indeed, when the user stands in front of an artwork, for example experiencing a tactile fruition of a 3D model, IMU measurements could be very noisy and lead to positioning errors.

PDR solutions can be executed also when the camera correctly captures the colored tapes. In this case, the measurements provided by the IMU system can be aggregated to the heading information provided by the computer vision algorithms, in order to improve the accuracy of the estimated user position along the path.

5.1 Activity Recognition

The *Activity Recognition* is devoted to classify the human motion. This step is crucial, since according to the output of this subsystem, different models are applied to track the displacement of the user. This step is also difficult to be performed when the number of motions to be identified is high, however, in this set up only a limited number of activities are required.

Most of the cultural sites, indeed, are organized in planar environments connected by stairs and/or elevators. The user walks in this environment and stops to appreciate the artworks. Using markers to identify the floor or the area, the user moves in a two-dimensional space and only 2 activities, i.e., *standing still* and *walking* need to be recognized. These two activities can be easily detected by exploiting the accelerations recorded by the IMU, as proposed in [24–26].

5.2 Heading Estimation

The *Heading Estimation* aims at computing the direction of the user when visiting the cultural sites. It is related only to the *walking* activity since when the user stops is supposed standing still in front of an artwork.

The heading estimate is calculated with respect to the fixed Cartesian reference frame (i.e., the *Navigation Frame*, NF) and it is performed using data from gyroscopes and accelerometer merged with the information provided by the camera in a two-step procedure:

- The *Attitude Estimation*;
- The *Heading Refinement*.

The *Attitude Estimation* exploits the measurements collected from gyroscopes, accelerometers and magnetometers to provide the attitude of the smartphone reference frame (i.e., the Body frame) with respect to the NF. The attitude is computed as quaternions and an Extended Kalman Filter is applied to merge the data from the different sensors as detailed in [26, 27]. Once the quaternions are updated, both the rotation matrix $\mathbf{R}_b^n$ from the Body frame to the NF and the heading γ_k with its uncertainty Γ_k can be easily retrieved. The initial condition q_0 is obtained from the acceleration and the data provided by the camera considering the user standing still when the system is activated as in [26].

Data from gyroscopes and accelerometers are affected by bias that downgrade the accuracy of the estimate. To reduce this effect, the bias is recalculated when the *standing still* activity is detected. Furthermore, the *Heading Refinement* uses the data from the camera to bound the drift error. Specifically, the only the heading γ_k and the related uncertainty Γ_k feed the correction step and the following simple update is performed

$$\phi_k = (1 - W_k)\gamma_k + W_k \gamma_{C,k} \tag{1}$$

where W_k is a gain computed according to EKF equation as

$$W_k = \frac{\Gamma_k}{\Gamma_k + R}$$

where R is the uncertainty related to the heading measurement $\gamma_{C,k}$ performed by processing data from the camera and is supposed known and time-invariant.

5.3 Position Estimation

The *Position Estimation* computes the position (x, y) of the user with respect to the *Navigation Frame* NF. To this end, the rotation matrix $\mathbf{R}_b^n$ provided by the Heading Estimation is used to project the accelerometer data in the the NF: in this way the acceleration along the z-axis lies on the sagittal plane of the user, perpendicular to the floor. This signal is fundamental during the *walking* activity, since it is analyzed to to identify the step event. Specifically, the local minima and the local maxima of the vertical acceleration signal is retrieved to perform both peak detection and zero crossing detection. The sharp changes to the vertical acceleration associated with the heel strike. These features are also exploited to compute the cardinality c_i of the set of samples to be processed to cope with different walking speed. During *walking* activity, the displacement l_i is estimated as proposed in [28].

The initial position is supposed known while the position (i.e., the position of the first marker met by the user when approaching an artwork) of the user during *walking* is recursively computed, by estimating the length of the stride on step event detection i

$$\begin{bmatrix} x_i \\ y_i \end{bmatrix} = \begin{bmatrix} x_{i-1} \\ y_{i-1} \end{bmatrix} + l_i \begin{bmatrix} \sin \bar{\phi}_i \\ \cos \bar{\phi}_i \end{bmatrix} \tag{2}$$

where l_i is the displacement of the user between two step events with respect to the NF and $\bar{\phi}_i$ is the average heading in the same time interval. Finally, the user displacement is $l_i = \begin{bmatrix} 0 \, 0 \end{bmatrix}^T$ when the output of the classification phase is the *standing still*.

6 Experimental Validation

To evaluate the performance of the proposed system, several experiments have been carried out. Specifically, two different types of trials have been considered. The first set aims at assessing the performance of the system comparing the tracking results with a ground truth. The second set is devoted to prove the effectiveness of the approach in a real scenario. In both cases a Samsung Galaxy S6 (SM-G920F), running Android 6.0.1 is used: it is equipped with an IMU-MPU6500 by Invensense and an IMX240 camera by Sony. Data from IMU are available at sampling frequency 100 Hz, while the images from the camera are acquired at 20 Hz.

In Table 1, the parameters used to initialize Tracking System (TS) during the experiments are shown.

Table 1 Parameters initialization

Parameter	Value
β	0.48
α_x	0.02
α_y	0.01
α_z	0.03

The parameter β is related to the human gait and has been calibrated on the user adopting the procedure introduced in [27]; the covariances associated to the accelerations $\{\alpha_x, \alpha_y, \alpha_z\}$ are used to discriminate between different human activities and are computed at the beginning of the experiment when the user is supposed standing still for 10 s.

To assess the performance of the system, the optical motion capture system OptiTrack has been used to compute the ground truth (GT). The system for motion capture exploits 10 infrared cameras to detect the position of markers in a limited area: it reach the accuracy of 10^{-4} m. To build the GT, the smartphone has been equipped with 4 markers: the CoM of the markers corresponds to the CoM of the smartphone. The accuracy of the proposed system is evaluated according to the estimated error on checkpoints. Specifically, the considered key performance indicator is represented by the Euclidean distance between the estimate (i.e., PDR or TS) and the corresponding points on the GT.

In the first trial, the user walks on an *S shaped* path, (10 m) long. Both the results obtained using PDR-only and the complete tracking system are compared with the GT and reported in Fig. 5. Considering PDR-only approach, the error accumulated is relevant, although the path is short and the opposite curves partially compensate the drift. Using the Tracking System, the error is reduced to 0.54 cm in the worst case.

The second experiment aims at evaluating the accuracy of the estimate when a closed loop is executed. To this end the user is required to repeat a square-path 5

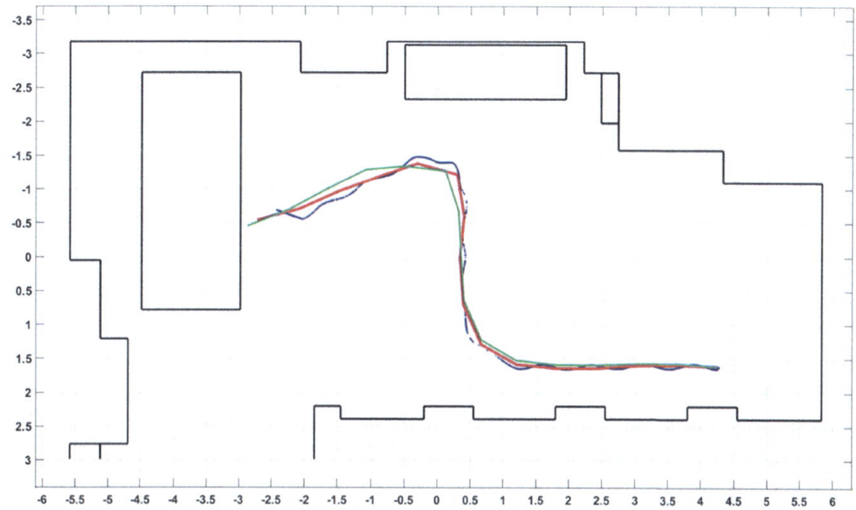

Fig. 5 Results of the S-path experiments: the $(x–y)$ axes represent the NF $[m]$

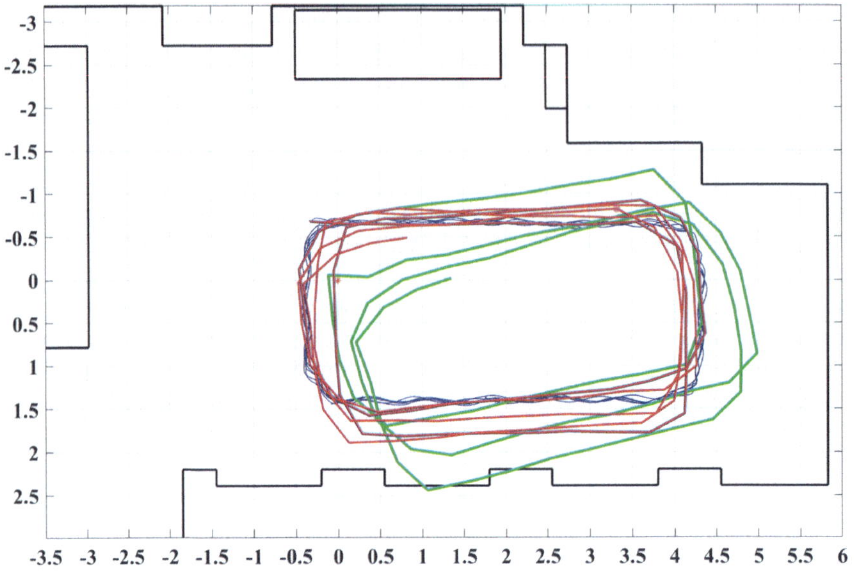

Fig. 6 Results of the Q-path experiments: the $(x-y)$ axes represent the NF $[m]$

Table 2 S and square-path: performance index

	Mean err $[m]$	Min err $[m]$	Max err $[m]$	Std dev $[m]$
S-path				
PDR	0.21	0.01	1.25	0.09
TS	0.10	0.001	0.54	0.02
Square-path				
PDR	0.66	0.15	1.77	0.22
TS	0.34	0.15	0.61	0.02

time without stops (see Fig. 6).The user walks for 130 m: since the user turns in the same direction, the shape of the path does not compensate the drift. The heading drift makes the PDR error unbounded, thus the covariance of the error is larger than the one obtained in the previous trial. When the heading is continuously corrected by the vision system the corresponding covariance does not change.

The values of the key performance indicator selected to asses the accuracy of the proposed system are collected in Table 2.

To prove the effectiveness of the approach in a real scenario, the ARIANNA system has been tested in the Farm Cultural Park, an open-air museum in city center of the Favara (AG), Italy. The project of the Farm Cultural Park has involved

Fig. 7 ARIANNA installation at Farm Cultural Park, Favara, Italy

several semi-abandoned buildings that were completely renovated. They host expositions, exhibitions by international and local artists. The system ARIANNA has been installed to help the visually impaired people visit the outdoor museum, as shown in Fig. 7.

In the trial, the user travels among the buildings, exploiting the path shown in Fig. 8. The path is about 76 m: in this case, the ground truth is not available, so the key performance indicator is represented by the error accumulated at the end of path. Considering PDR, the final positioning error is 3.1 m, that represents about 4% of the distance traveled. This error is limited, since the path is almost straight, however, using the correction provided by the camera, the error is reduced to 0.41 m, that represents less than the 1% of all the distance traveled.

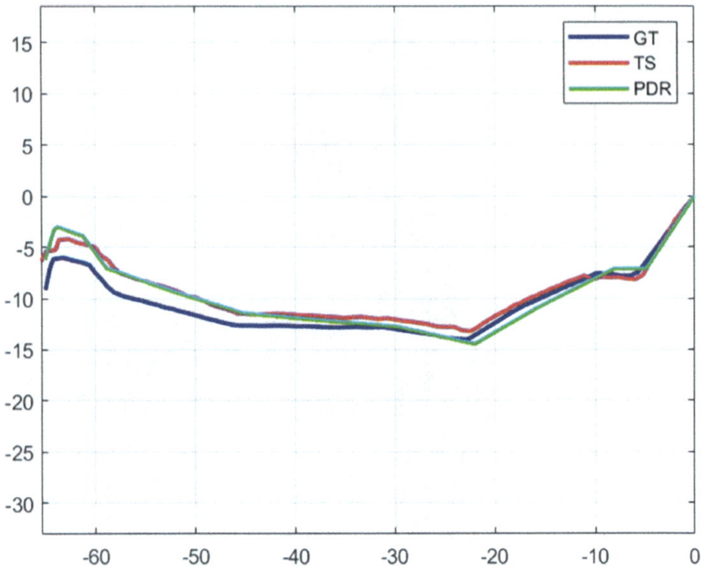

Fig. 8 Real and estimated path in a real installation: the $(x\text{–}y)$ axes represent the NF $[m]$, GT (blue line), PDR (green line), and TS (red line)

7 Conclusions

In this chapter we have presented ARIANNA, an innovative smartphone-centric tracking system for indoor and outdoor environments, based on the joint utilization of dead-reckoning and computer vision techniques. The system is explicitly designed for visually impaired people and has been tested in collaboration with key institutions such as the Italian Blind Association (Unione Italiana Ciechi). Many have been the installations (permanent or temporary) that have been provided to the users to be tested. In the occasion of these events, we interviewed 20 visual impaired people (10 low vision users and 10 blind users), asking them to answer to an assessment questionnaire. In both cases, the answers where very encouraging, with a satisfaction grade of 90%. A summary of the users' feedback is provided in Tables 3 and 4.

In real outdoor tests, such as the one carried in the installation present in the Farm Cultural Park, the system is able to suitably reduce the navigation error with respect to the results obtained using PDR only, as demonstrated by the trial carried out using the precise optical tracking system as GT. Future development of the system will include also a model of the hand movement and possibly the design of a vibration feedback to tell the user how to correct its location. Based on the presented tests and experiments, we have implemented and embedded all the algorithms into an app available for both Android and iOS platforms.

Table 3 Answers by low vision people

Question		Answer			
		Yes	No		
1	Have you understood the shape of the path?	9	1		
		None	Mild	Moderate	Very
2	Have you found difficult the change in direction?	2	2	6	0
3	How easy was to follow the path?	0	2	6	2
4	Do you need a learning phase?	1	3	4	2
5	The vibration is useful?	0	0	4	6
6	Have you noticed delays?	6	2	1	1
		Unusable	Not good	Satisfactory	Good
7	What's about the haptic interface operation?	0	3	4	3
		None	Mild	Moderate	Very
8	Do you think this app can substitute tactile paving?	0	1	2	6
9	Do you think interactive information are useful?	0	1	3	6
10	Do you think ARIANNA will increase your independence?	0	0	4	6
		With direction	Without direction	Both	
11	Which operaton mode do you prefer?	6	2	2	
		Indoor	Outdoor		
12	In which context is ARIANNA useful?	8	9		
		None	Mild	Moderate	Very
13	Have you found difficulties in the use of ARIANNA?	0	0	6	3

Table 4 Answers by blind people

Question		Answer			
		Yes	No		
1	Have you understood the shape of the path?	6	4		
		None	Mild	Moderate	Very
2	Have you found difficult the change in direction?	3	6	1	0
3	How easy was to follow the path?	0	4	6	1
4	Do you need a learning phase?	0	3	4	2
5	The vibration is useful?	0	1	6	3
6	Have you noticed delays?	6	3	1	0
		Unusable	Not good	Satisfactory	Good
7	What's about the haptic interface operation?	0	1	7	2
		None	Mild	Moderate	Very
8	Do you think this app can substitute tactile paving?	0	0	2	8
9	Do you think interactive information are useful?	2	1	2	5
10	Do you think ARIANNA will increase your independence?	1	0	6	3
		With direction	Without direction	Both	
11	Which operaton mode do you prefer?	3	5	2	
		Indoor	Outdoor		
12	In which context is ARIANNA useful?	5	8		
		None	Mild	Moderate	Very
13	Have you found difficulties in the use of ARIANNA?	4	6	0	0

Acknowledgements This work was supported in part by the European Union (EU) Funded Research Projects INTERREG V-A Italia-Malta 2014–2020 "I-ACCESS: Implementing the Accessibility to Urban Historic Center's Use and Knowledge", project number C1-1.1–41. The project is carried out by the DARCH department at University of Palermo, Italy.

References

1. WHO, Visual impairment and blindness (2014), http://www.who.int/mediacentre/factsheets/fs282/en/. Accessed Oct 2018
2. WHO, World sight day: 13 October 2011 (2011), http://applications.emro.who.int/dsaf/. Accessed Oct 2018
3. F. Felice, F. Renna, C. Attolico, A. Distante, Omero: a multimodal system that improves access to cultural heritage by visually impaired people. Archeologia e Calcolatori supplemento 1 (2007)
4. V. Cantoni et al., A multi-sensory approach to cultural heritage: the Battle of Pavia exhibition. IOP Conf. Ser.: Mater. Sci. Eng. **364**, 012039 (2018)
5. "Building tourism markets accessible to vision impaired-people", https://www2.ift.edu.mo/NewsPortal/building-tourism-markets-accessible-to-vision-impaired-people/. Accessed Oct 2018
6. D. Croce, P. Gallo, D. Garlisi, L. Giarré, S. Mangione, I. Tinnirello, ARIANNA: a smartphone-based navigation system with human in the loop, in *22nd Mediterranean Conference of Control and Automation (MED)* (2014), pp. 8–13
7. D. Croce, L. Giarré, F.G. La Rosa, E. Montana, I. Tinnirello, Enhancing tracking performance in a smartphone-based navigation system for visually impaired people, in *24th Mediterranean Conference of Control and Automation (MED)* (2016)
8. Italian Patent N. BG2014A000054, Sistema di navigazione per non vedenti, presented 2015, patented 2017
9. D. Croce, L. Giarré, F. Pascucci, I. Tinnirello, G. Galioto, D. Garlisi, A.L. Valvo, An indoor and outdoor navigation system for visually impaired people. IEEE Access **7**, 170406–170418 (2019)
10. R. Velázquez, Wearable assistive devices for the blind, *Smart Environment: Issues and Characterization*. LNEE, vol. 75 (Springer, Berlin, 2010), pp. 331–349
11. N. Fallah, I. Apostolopoulos, K. Bekris, E. Folmer, The user as a sensor: navigating users with visual impairments in indoor spaces using tactile landmarks, in *Proceedings of the 2012 ACM Annual Conference on Human Factors in Computing Systems (CHI'12), Austin, Texas* (2012), pp. 425–432
12. B. Krach, P. Roberston, Cascaded estimation architecture for integration of foot-mounted inertial sensors, in *IEEE/ION Position, Location and Navigation Symposium, Monterey, CA* (2008), pp. 112–119
13. K. Kaemarungsi, P. Krishnamurthy, Properties of indoor received signal strength for WLAN location fingerprinting, in *Proceedings of 1st Annual International, (MobiQuitous04)* (2004)
14. J.A.M. Ladd, K.E. Bekris, A.P. Rudys, D.S. Wallach, L.E. Kavraki, On the feasibility of using wireless ethernet for indoor localization. IEEE Trans. Wirel. Commun. **5**(10), 555–559 (2006)
15. A.R. Jimènez, F. Seco, J.C. Prieto, J. Guevara Rosas, Indoor pedestrian navigation using an INS/EKF framework for yaw drift reduction and a foot-mounted IMU, *Workshop on Positioning Navigation and Communication, Dresden* (2010), pp. 135–143
16. A.R. Jimènez, F. Seco, F. Zampella, J.C. Prieto, J. Guevara Rosas, Improved heuristic drift elimination (iHDE) for pedestrian navigation in complex buildings, in *International Conference on Indoor Positioning and Indoor Navigation, Guimaraes* (2011), pp. 1–8
17. L. Filardo, F. Inderst, F. Pascucci, C-IPS: a smartphone based indoor positioning system, in *International Conference on Indoor Positioning and Indoor Navigation (IPIN), 4–7 October 2016, Alcala de Henares, Madrid, Spain* (2016)
18. V. Singh, R. Paul, D. Mehra, A. Gupta, V.D. Sharma, S. Jain, C. Agarwal, A. Garg, S.S. Gujral, M. Balakrishnan, K. Paul, P.V.M. Rao, D. Manocha, Smart'Cane for the visually impaired: design and controlled field testing of an affordable obstacle detection system, in *TRANSED 2010: 12th International Conference on Mobility and Transport for Elderly and Disabled Persons*, vol. 53 (2010), pp. 1689–1699
19. B. Hoyle, D. Waters, Mobility AT: the batcane (ultracane), in *Assistive Technology for Visually Impaired and Blind People* (Springer, Berlin, 2008), pp. 209–229

20. A. Wachaja, P. Agarwal, M. Zink, M.R. Adame, K. Moller, W. Burgard, Navigating blind people with a smart walker, in *2015 IEEE/RSJ International Conference on Intelligent Robots and Systems (IROS), Hamburg* (2015), pp. 6014–6019
21. A. Cosgun, E.A. Sisbot, H. Christensen, Guidance for human navigation using a vibro-tactile belt interface and robot-like motion planning, in *International Conference on Robotics and Automation* (2014)
22. M. Pielot, R. de Oliveira, *Peripheral Vibro-Tactile Displays, Mobile HCI 2013 - Tactile User Interfaces* (ACM Press, New York, 2013). https://doi.org/10.1145/2493190.2493197
23. M. Pielot, How the phone's vibration alarm can help to save battery, http://pielot.org/2012/12/11/how-the-phones-vibration-alarm-can-help-to-save-battery/. Accessed 12 May 2014
24. G. Galioto, I. Tinnirello, D. Croce, F. Inderst, F. Pascucci, L. Giarré, Demo: sensor fusion localization and navigation for visually impaired people, in *MobiCom* (2017)
25. G. Galioto, I. Tinnirello, D. Croce, F. Inderst, F. Pascucci, L. Giarré, Sensor fusion localization and navigation for visually impaired people, in *ECC* (2018)
26. F. De Cillis, L. Faramondi, F. Inderst, S. Marsella, M. Marzoli, F. Pascucci, S. Setola, Hybrid indoor positioning system for first responders, in *IEEE Transactions on Systems, Man, and Cybernetics: Systems*, vol. 50(2) (2020), pp. 468–479
27. L. Faramondi, F. Inderst, F. Pascucci, R. Setola, U. Delprato, An enhanced indoor positioning system for first responders, in *International Conference on Indoor Positioning and Indoor Navigation* (2013)
28. H. Wienberg, Using the ADXL202 in pedometer and personal navigation applications, Analog devices AN-602 application note (2002)

A Robotic Humanoid for Information Exploration on Cultural Heritage Domain

Agnese Augello, Giovanni Pilato, Gaia Trecarichi, Giorgio Vassallo and Filippo Vella

Abstract The work presented here illustrates an humanoid robot capable of interacting with an human user within the Cultural Heritage domain. Two different and complementary AI approaches, namely sub-symbolic and symbolic, have been implemented and combined together to design the framework of a robot having both rational and intuitive capabilities. Furthermore, the robot is capable of providing information expressively and of adapting its behavior according to the emotional content of the artworks descriptions. This could make the robot more effective in providing information and entertaining the users.

1 Introduction

The significant development of technologies in Robotics and Artificial Intelligence can be profitably used to develop a set of new systems and installations able to engage the visitors with Cultural Heritage contents. All these systems, having as principal user human beings, should be oriented to tackle multimodal interactions with the users, exploiting different means for providing the required information.

A. Augello · G. Pilato (✉) · F. Vella
ICAR-CNR, Palermo, Italy
e-mail: giovanni.pilato@cnr.it

A. Augello
e-mail: agnese.augello@cnr.it

F. Vella
e-mail: filippo.vella@cnr.it

G. Trecarichi
Istituto Tecnico Tecnologico "Buonarroti", Trento, Italy
e-mail: gaia.trecarichi@scuole.provincia.tn.it

G. Vassallo
DIID-University of Palermo, Palermo, Italy
e-mail: giorgio.vassallo@unipa.it

D. Seychell and A. Dingli (eds.), *Rediscovering Heritage Through Technology*, Studies in Computational Intelligence 859,
https://doi.org/10.1007/978-3-030-36107-5_3

In the last years, many efforts have been conducted for improving the interaction with the user [32, 33] in order to provide intelligent systems capable of facilitating and encouraging the interest towards cultural attractions as well as supporting education and information diffusion. Many technologies have been introduced in museums aimed at the fruition of cultural heritage, and an overview is given in the related works section.

A solution that is appreciated by people of different ages is the possibility of interacting with a robot that can constitute an amusing and catching interface beyond traditional approaches.

A robot is typically used when a mechanical substitute of a real person is advisable. Such a system should be able to receive and understand what people are asking; moreover, it should propose contents that meet the requests and the expectations of the user, also showing an engaging behavior.

To reach this goal, capabilities of natural language understanding, processing of the information in the request, selecting the best content to provide and proposing the content with the most adequate and attractive manner, are required.

This chapter illustrates a robotic humanoid framework able to perform information exploration in the cultural heritage domain.

The robot is capable of interacting naturally with the user. It is equipped with cognitive-oriented approaches, combined with symbolic ones so that it can show an "intuitive" association capability to convey information to the user. Furthermore, this capability is enhanced by a text-based emotion detection, which makes it possible for the robot to provide the information to the user in an engaging and possibly amusing manner.

The intuitive/associative capability has been obtained using the LSA (Latent Semantic Analysis) methodology [18, 30]. The approach lets arise the latent relations between the words belonging to an extensive collection of documents; a semantic space is then built to map words as vectors so that the semantic similarity between words can be calculated using the cosine similarity measure between vectors in that space.

The robot exploits LSA-enhanced ontologies for an information exploration system on the cultural heritage domain. The robot is supplied with logical reasoning capabilities arising from a formal approach and with the ability to extract semantics from sentences and concepts so that associations between concepts are "more intuitive than logic". In particular, the "conceptual space" has been created starting from a large text corpus dealing with a specific domain.

Furthermore, the robot can read information about content with an emotional enrichment of the text. All the time, it can associate an emerging emotion to the text it can show this emotion through its behavior: changing its position, varying the movements of its arms and modifying the tone of its voice.

The chosen application domain concerns the cultural heritage of museums, namely, the one related to the Rijksmuseum of Amsterdam. We took this choice because of the vast quantity of documents available on the Rijksmuseum website, which is quite well structured and easy to explore automatically.

The next section gives an overview of related works on the topic; Sect. 3 deals with the interaction system of the humanoid robot; Sect. 4 illustrates the process of construction of data-driven semantic spaces that can be interpreted somehow as "conceptual" spaces; Sect. 5 outlines the process of emotion detection from texts, which exploits data-driven conceptual spaces, Sect. 6 illustrates the knowledge base of the robot; Sect. 7 deals with the description of the dialogue engine and illustrated the communicative Skills of the robot as well as its expressive capabilities; Sect. 8 gives a domain application and a sample of interaction; conclusions are finally given.

2 Related Works

The adoption of technologies in museums for the fruition of cultural heritage is not new. Some attempts have been made for enriching the user experience improving the ways a visitor can approach the artistic content. According to [13], robots are used in a museum with three modalities: museum guide, telepresence, and installation. The robotic museum guide is the most common robotic installation in a museum. The robot moves in the rooms of the museum avoiding obstacles and, in the meantime, verbally interacting with visitors to describe artistic contents.

Going some year in the past, in 1998, the robot Chips-Sage has been introduced at the Carnegie Museum of Natural History in Pittsburgh, to accompany visitors in the Dinosaur Hall and to provide audiovisual information in the form of a unidirectional narrative speech [22, 40]. The Smithsonian National Museum of American History introduced the robot Minerva, which offered information to visitors [37]. This robot was able to produce facial expressions and communicate its emotional state, following people behavior. The shown emotions improved the engagement capability of the robot and its effectiveness in the guidance [37].

A robot with interaction capabilities has been tested at the Archaeological Museum of Agrigento, Italy, within the CiceRobot project. The robot allowed visitors to make questions that were filtered through a semantic module [20]. In 2000, a group of three different robots had been introduced at the Museum für Kommunikation in Berlin. These robots had to accomplish different tasks, such as instruction and entertainment. The instructive robot was a museum guide that accompanied visitors giving explanations about the exhibit. It was able to indicate which object it was referring to and could show additional contents on a screen [9]. Another interesting example of a guide robot is given by Urbano. It was employed in an exhibition in Spain, and it was able to show emotional behavior. In particular, it could behave according to what was done by the visitors [3].

Also, the famous robot Asimo, produced by Honda, has been employed as a museum guide at the Japan National Museum of Emerging Science and Innovation. The general activity was to guide a group of people with a focus on the interaction with them. It required people to raise their hand to have the possibility to talk to it and ask questions [11].

Another segment where robots are employed, within the same field, is telepresence. Robots are connected through the internet, allowing visitors to explore the museum remotely. In general, all these mobile robotic platforms are equipped with a screen and a camera. Some examples are the robot Csiro [34] and the robot Norio [21].

The robot Csiro has been employed at the National Museum of Australia to make people, navigate through the museum item even if they live in inner or far areas of Australia [34]. The robot Norio, produced by Droids Company, allowed people with impaired mobility to visit the National Centre for Monuments, Château d'Oiron, France. At the ground floor of the museum, a maneuvering station let the visitor remotely drive the robot, located on the first-floor [10, 21].

An example of artistic installation involving a robot is provided by the robot Paul, conceived by the artist and scientist Patrick Tresset. This installation consists of a robotic arm formed by a left-hand robotic arm holding a black biro and a pan-tilt camera. It can draw a portrait of the person that is sat in front of it. It has been exhibited for the first time at the Tenderpixel Gallery in London in 2011 [38].

3 The Humanoid Robot Interaction System

The proposed architecture is integrated with and makes use of a Pepper humanoid robot [24]. This robot is built to play the role of a companion with people so that it can communicate in a natural and intuitive manner, making use of its body movements and its voice. It can move in all directions; it can change the position of its head, arms, and back. It has ultrasound transmitters and receivers, laser sensors, and obstacle detectors that make it able to move to an indoor environment without difficulties. Moreover, Pepper has tactile sensors in its hands, that can be exploited in order to make its interaction even more natural, realistic, and engaging [24]. All these characteristic can be exploited for social interaction, making this humanoid robot a suitable interface for a cultural heritage narrator.

In the approach illustrated in this paper, the robot exploits its Internet connection capability to interface itself with a family of "ad hoc" built modules retrieving the most appropriate information that is not present into its internal Knowledge base.

The approach allows the robot to manage the activity with the users by verbally interacting with them, while adapting its expressions according to the semantic and emotional contents of the artwork descriptions or, more generally, of the information to be provided. The robot is endowed with knowledge about a case-study museum and the artworks contained in it. This knowledge can be queried by a conversational engine that is formalized in different ways: a pre-built set of question-answers modules, a domain-specific ontology, formalized in Cyc [23] and an LSA sub symbolic semantic space [1, 18, 26], where both the question-answers modules and ontology concepts are mapped. This choice makes it possible to retrieve information when a formal, rule-based reasoning path, cannot be found. Besides, a retrieval module

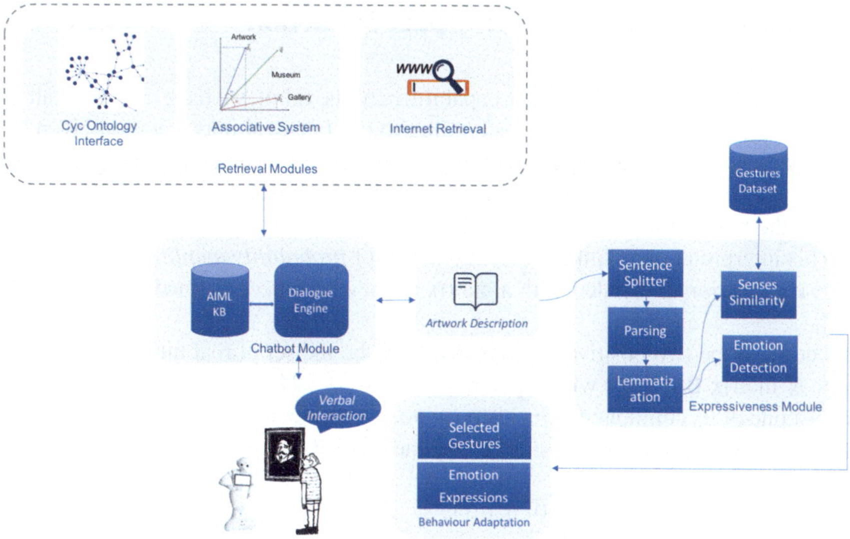

Fig. 1 System infrastructure

allows the robot to query the Internet in order to find information that has not been formalized into the system. Figure 1 shows the main components of the system:

- The chatbot conversational Engine and the AIML Kb (the set of question/answers modules);
- The retrieval modules:
  - the module querying the Cyc Ontology
  - the associative system querying the semantic space;
  - Internet retrieval module;
- The expressiveness module used to manage the expressive exposition of the artworks descriptions, relying on:
  - a natural language processing module;
  - an emotion detection module;

The right combination and integration of the components mentioned above makes it possible having a system able to retrieve information from multiple sources and to describe the content in natural language, with an emphasis on emotion relevant information. The details are given in the following sections.

4 Data-driven "Conceptual" Space Induction

To emulate the associative/intuitive capabilities of the robot, we have used the Latent Semantic Analysis paradigm [18] according to its statistical interpretation given by [30], where the T-SVD algorithm, as used in the Latent Semantic Analysis paradigm, acts as an *estimator*, conveying statistically significant information from the sample data to the inferred model.

This interpretation is linked to the concepts of *probability amplitude* and *probability distribution* associated with a matrix as they have been defined in Pilato et al. [30].

Let M, N be two positive integers and let $\mathbb{R}$ be the set of real numbers. Given a $M \times N$ matrix $\mathbf{B} = [b_{ij}]$ with $b_{ij} \in \mathbb{R}$, $i \in [1, 2, \ldots, M]$, $j \in [1, 2, \ldots, N]$ where at least one of its components $[b_{ij}]$ is positive, we define a set J, composed of all the pairs (i, j) that identify the positive components of $\mathbf{B}$, i.e.:

$$J = \{(i, j) : b_{ij} > 0\} \;\; i \in [1, 2, \ldots, M], j \in [1, 2, \ldots, N] \tag{1}$$

The *probability amplitude* associated with $\mathbf{B}$, is defined as the $M \times N$ matrix $\Psi = [\psi_{ij}]$ resulting from the mapping $p_a(\cdot)$:

$$\Psi \equiv p_a(\mathbf{B}) : \mathbb{R}^{M \times N} \rightarrow [0, 1]^{M \times N} \tag{2}$$

whose elements $[\psi_{ij}]$ are computed as:

$$\psi_{ij} = \begin{cases} \frac{b_{ij}}{\sqrt{\sum_{(i,j)\in J} b_{ij}^2}} & if \;\; b_{ij} > 0 \\ 0 & if \;\; b_{ij} \leq 0 \end{cases} \tag{3}$$

so that $\forall (i, j)$ it is $\psi_{ij} \geq 0$ and $\sum_{i=1}^{M} \sum_{j=1}^{N} \psi_{ij}^2 = 1$.

The *probability distribution* associated with a matrix $\mathbf{B}$ the $M \times N$ matrix resulting from the mapping $p_d(\cdot)$ is defined as:

$$\mathbf{B}^{(2)} \equiv p_d(\mathbf{B}) : \mathbb{R}^{M \times N} \rightarrow \mathbb{R}^{M \times N} \tag{4}$$

whose elements are the squares of the elements of $\mathbf{B}$, i.e. $\mathbf{B}^{(2)} = [b_{ij}^2]$. The method starts with a raw data matrix $\mathbf{A}$ consisting of positive values. In our study the raw data matrix $\mathbf{A}$ is the term-document co-occurrence matrix. From $\mathbf{A}$ a real-valued normalized matrix $\mathbf{Q}$ is computed by dividing every element for the sum of all elements of $\mathbf{A}$.

$$\mathbf{Q} = \frac{\mathbf{A}}{\sum \sum a_{ij}}, \forall a_{ij} \epsilon \mathbf{A}. \tag{5}$$

If we call Ψ_Q the matrix:

$$\Psi_Q = \left[\sqrt{q_{ij}}\right] \tag{6}$$

The matrix Ψ_Q can be decomposed with the SVD technique:

$$\Psi_Q = \mathbf{U}\Sigma\mathbf{V}^T \tag{7}$$

and its best *rank-r* decomposition $\Xi = [\xi_{ij}]$ is obtained by applying the T-SVD technique, which minimizes the Frobenius distance $d_F(\Xi, \Psi_Q)$, given r:

$$\Xi = \mathbf{U}_r\Sigma_r\mathbf{V}_r^T \tag{8}$$

Even if Ξ is not a probability distribution, the computation of Ξ makes it possible to identify, without any further addition of external information, the probability distribution we are looking for. As shown in Pilato et al. [30], it theoretically suffices computing the probability amplitude associated to Ξ, i.e. $p_a(\Xi)$, and consequently calculating the probability distribution $p_d(p_a(\Xi))$ associated to $p_a(\Xi)$. The aforementioned Frobenius distance $d_F(\Xi, \Psi_Q)$ constitutes an upper bound to the Hellinger distance between the sample probability $\mathbf{Q}$ and the probability distribution estimated by the procedure.

4.1 Mapping New Elements into the "Conceptual" Space

The matrices Σ_r and the $\mathbf{U}_r$ provided by LSA can be used for computing the vector encoding of new documents into the data-driven conceptual space [8, 18, 29].

A document d is sub-symbolically represented by a $dim(V)$-dimensional word occurrence vector $\mathbf{d}$, from which it is computed a vector $\mathbf{q}$ as:

$$\mathbf{q} = \sqrt{\frac{\mathbf{d}}{\sum_i\sum_j a_{ij}}} \quad \forall i \in [1, dim(V)] \tag{9}$$

where a_{ij} is the generic (i, j) element of the matrix $\mathbf{A}$. The r-dimensional vector $\mathbf{d}_r$ representing projection of d in the induced conceptual space is then obtained from the vector $\mathbf{q}$ by means of the following mapping formula:

$$\mathbf{d}_r = \mathbf{q}^T\mathbf{U}_r\Sigma_r^{-1} \tag{10}$$

5 Emotion Detection from Text

The sub-symbolic conceptual space is also exploited to implement the emotion detection capabilities of the robot. In particular, it deals with the detection of emotions in sentences. These capabilities are used for expressing in a natural and emotion enriched manner the information provided to the user [6].

The six Ekman basic emotions, *anger*, *disgust*, *fear*, *joy*, *sadness* and *surprise*, have been used to implement this feature. We have exploited the procedure that has been introduced and applied in [8, 29].

In particular, we have six sets E_{anger}, $E_{disgust}$, ..., $E_{surprise}$ of vectors constituting the sub-symbolic coding of specific subsets of words identifying a basic emotion. The generic vector belonging to one of the sets is denoted as $\mathbf{b}_i^{(e)}$ where $e \in \{$"$anger$", "$disgust$", "$fear$", "joy", "$sadness$", "$surprise$"$\}$ and i is the index that identifies the ith $\mathbf{b}_i^{(e)}$ in the e set. Specifically, $\mathbf{b}_i^{(e)}$ is computed by applying formula (10). Analogously, any sentence s can be mapped into the Data Driven "conceptual" space by computing a vector $\mathbf{s}_R$, by using the same procedure. Then, it is possible to compute the emotional characterization of the sentence by exploiting the vectors $\mathbf{b}_i^{(e)}$; in particular, fixed $\mathbf{s}_R$, for each set E_e it is computed the weight:

$$w_e = \max cos(\mathbf{s}_R, \mathbf{b}_i^{(e)}) \tag{11}$$

once all the six w_e weights are computed, the vector $\mathbf{f}_s$, associated to the vector $\mathbf{s}_R$, and by consequence to the sentence s, is calculated as:

$$\mathbf{f}_s = \left[\frac{w_{(anger)}}{\sqrt{\sum_e w_e^2}}, \frac{w_{(fear)}}{\sqrt{\sum_e w_e^2}}, \cdots, \frac{w_{(surprise)}}{\sqrt{\sum_e w_e^2}} \right] \tag{12}$$

The vector $\mathbf{f}_s$ finally constitutes the *emotional fingerprint* of the text chunk s in the emotional space, which is a six-dimensional hypersphere where all the sentences are mapped, and each region of this hypersphere is associated to a set of emotional manifestation of the robot.

The element of $\mathbf{f}_s$ having the highest value determines the main emotion expressed by the sentence. A minimum value of threshold Th_e is experimentally fixed in order to determine neutral sentences.

6 The Knowledge Base of the Robot

The knowledge of the robot consists of three main components:

- a data-driven "conceptual" space created through LSA;
- a Cyc-based ontology representing a specific-domain knowledge;
- a geometric representation of the ontology obtained by mapping ontological concepts in the data-driven "conceptual" space: this mapping characterizes the "intuitive/associative" capabilities of the robot.

The chosen application domain is the cultural heritage of museums, in particular, the one of the Rijks Museum of Amsterdam. We have chosen this museum because of the great quantity of online information that is possible to obtain about the collections that are present there.

6.1 Ontology by CYC Knowledge Base

The Cyc knowledge base (KB) at present is claimed to be one of the largest and most complete general KB equipped with a good performing inference engine. The CYC technology has been preferred to others because of its suitableness for common sense representation [23]. Such a common sense, together with a more abstract ontology, permits to extend inferences to more generic facts. Cyc is suitable for automated logical inference to support knowledge-based reasoning applications [23]. In Cyc, the knowledge base is composed of Microtheories (Mt), that are defined as specific collections of concepts and facts regarding a specific domain.

Since Cyc organizes concepts and assertions into Microtheories, a new micro-theory can be created in order to represent the knowledge about the involved domain, which has not yet been included in the standard KB.

A microtheory contains facts related to a specific domain, cyc-collections defining new and specific concepts, new predicates, assertions, and inference rules. Micro-theories are linked to some of the existing ones so that every assertion present in them is still valid in the new micro-theories, this criterion has the consequence of establishing a hierarchy between micro-theories.

In this specific case, we have extended Cyc with the knowledge about the robot itself and with the specific information about cultural heritage domain chosen as a case of study: the Rijks museum of Amsterdam.

6.1.1 Pepper Robot Self-knowledge

Some information regarding the structure of the Pepper robot has been coded into an ontology that we have implemented into the Cyc Commonsense Knowledge Base, with a similar approach of what illustrated in [16].

As an example, some of the most common predicates and concepts that we have used, among the many available ResearchCyc predefined items, are illustrated below:

- *#$genls* predicate: it relates a given collection to those collections that subsume it;
- *#$isa* predicate: it relates things of any kind to collections of which they are instances;
- *#$managesToPerformActOfType*: it is an instance of BinaryPredicate and FirstOrderCollectionPredicate. the assertion (*managesToPerformActOfType AGENT ACT-TYPE*) means that AGENT manages to perform an instance of ACT-TYPE.

- *#$properPhysicalPartTypes*: it is a RuleMacroPredicate and a specialization of physicalPartTypes. (*properPhysicalPartTypes TYPE1 TYPE2*) means that every instance of TYPE1 has an instance of TYPE2 as a proper physical part;
- *#$Robot*: it is a device which operates autonomously, moving about or manipulating physical objects;
- *#$RobotHand*: it is the collection of mechanical devices that can manipulate objects and/or materials automatically.

In what follows, we report a sample of constants and predicates that we have defined and inserted into the ResearchCyc KB:

- *#$Pepper-TheRobot*: it is the collection of the Pepper robots;
- *#$IcarPepper*: it is an instance of #$Pepper-TheRobot and is the robot that we have used for the experiments;

Some of the assertions that we have inserted into the ResearchCyc KB are:

- (*#$genls #$Pepper-TheRobot #$Robot*) : #$Pepper-TheRobot is under the more comprehensive category of the Robots;
- (*#$isa #$PepperNao #$Nao-TheRobot*) #$PepperNao is an instance of the collection #$Pepper-TheRobot;
- (*#$properPhysicalPartTypes #$Pepper-TheRobot #$RobotHand*) every instance of #$Pepper-TheRobot has an instance of #$RobotHand as a proper physical part;

A typical query to the inferential engine is:

- question: (*?X #$Robot ?Y*);
- answer:
  (#$properPhysicalPartTypes #$Robot #$RobotFoot) in BaseKB
  (#$genls #$Robot #$ElectronicDevice) in UniversalVocabularyMt
  (#$comment #$Robot "A device which operates autonomously, moving about or manipulating physical objects,") in UniversalVocabularyMt
  (#$isa #$Robot #$SpatiallyDisjointObjectType) in UniversalVocabularyMt
  ...

6.1.2 Rijks Museum Cyc Knowledge

We have used a previously developed Cyc micreotheory on the Rijks museum, named *Rijksmuseum-Mt*, containing facts related to the museum, cyc-collection defining new and specific concepts, new predicates, assertions, and inference rules [26]. The microtheories that constitute a generalization of the Rijksmuseum-Mt are: *GeopoliticalEntityVocabularyMt*, *FixedStructureVocabularyMt*, *HumanSocialLifeMt*, *UniversalVocabularyMt*, *ArtifactGVocabularyMt*, *BaseKB*. The RijksmuseumMt is linked to the other ones that we have mentioned through the *genlMt* predicate. Figure 2 illustrates the links of the created microtheory with the pre-existing microtheories of Cyc.

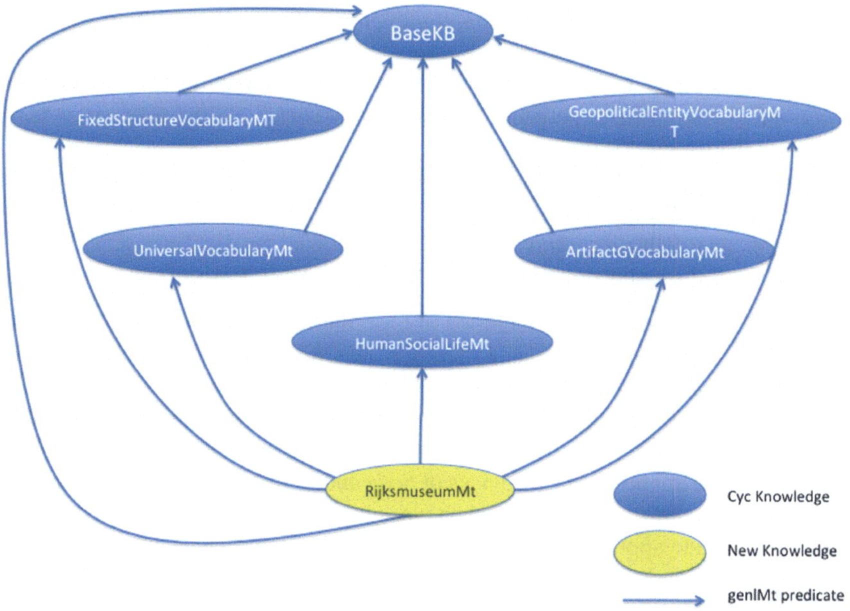

Fig. 2 RijksmuseumMt and its linking with the other pre-existing microtheories of Cyc

The collections used to represent the concepts associated with the museum are:

- *Museum*: it is the collection that represents the gathering of all museums. An instance of this collection is Rijksmuseum, which represents the museum hosting the pieces on which the robot is focused. It is a specialization of *ConstructionArtifact* defined in *FixedStructureVocabularyMt*.
- *Style*: it is the collection created to represent the artistic periods and the trends related to the hosted items. It is a specialization of *TemporalThing* defined in *BaseKB*. The instances of Style are: *Classicism*, *Romanticism*, *Impressionism*, *Gothic*, *Baroque*, *Mannerism*, *Renaissance*, *Humanism*.
- *ArtOnPaper*: this collection is a specialization of *ArtObject* defined in *ArtifactGVocabularyMt*. It includes all the art objects made of paper material, for example, prints, photographs, papyrus. Examples of instances are: *Krishna kills Shrigala, Tiger*.
- *Painting*: this collection is a specialization of *ArtObject* defined in *ArtifactGVocabularyMt*. It includes all the objects of art identified as paintings or similar, for example, the frescoes. Examples of instances are *ItalianLandscape*, *TheBattleOfLivorno*, *TheGust*, *LaCorniche*, *DucksNests*, *WindmillOnAPolderWate*.
- *Landscape*: this collection is a specialization of *Theme*, defined in *UniveralVocabularyMt*. It represents a particular subject, or theme, linked to the painting, that is a landscape that could be a rural landscape, maritime, etc. The instances of

this collection are: *DutchLandscape*, *ItalianateLandscape*, *Riverscape*, *Seascape*, *BeachView*.
- *People*: this collection is a specialization of Theme defined in *UniversalVocabularyMt*. It is linked to people. Examples of instances are the following: *Man*, *Woman*, *Child*.
- *Animals*: it is related to the animals' concept. Examples of instances are *Horse*, *WildAnimals*, *PoultryAndBirds*.
- *StillLife*: this collection represents the theme linked to still lifes. Examples of instances of this collection are: *StillLifeWithFood*, *StillLifeWithFlowers*.

Among the default collections included in Cyc, the KB uses: *Sculpture*, *ConstructionArtifact*, *SpatialThingLocalized*, *Artifact*, *ArtObject*, *Artist*, *ThemeCity*, *CapitalOfRegion*, *CalendarHalfCentury*, *CalendarCentury*.

Moreover, we have defined a set of predicates useful for relating the defined collections of RijksmuseumMt. In particular, some of them are: *hasBeenDevelopedIn*, a binary predicate applied to the instances of *Style* and *GeopoliticalEntity* used to state that a certain artistic movement has taken place in a certain region or city; *hasTheme*, a binary predicate applied to the instances of *ArtObject* and *Theme* used to specify the subject of a work of art. Other *ad-hoc* predicates are *comesFrom*, *worksDuring*, *cityOf*, *locatedIn*.

Furthermore, some of the predefined predicates have been used like: *operatesInRegion*, *conceptuallyRelated*, *comment*, *createdBy*, *artisticWorksCreated*, *capitalCity*, *startingDate*, *endingDate*, *operatesInRegion*.

Besides, a set of specific inference rules have been added to the ontology. As an example, the rule:

```
(implies
    (and    (isa ?OBJECT ArtObject)
            (createdBy ?OBJECT ?ARTIST)
            (comesFrom ?OBJECT ?LOCATION))
    (operatesInRegion ?ARTIST ?LOCATION))
```

means that if an *ArtObject* was created by a certain artist and it comes from a certain region, then the artist worked in that region.

6.2 Mapping the Ontology in the Conceptual Space

The semantic space which bridges the symbolic and the sub-symbolic knowledge of the robot has been created starting from a set of documents describing the museum, the collections hosted in it, the biography of the artists and the individual works [26].

A set of 5132 documents has been downloaded from the internet, they have been used to construct the "conceptual" space.

In order to give a sub-symbolic representation to the ontology, every newly created concept is mapped with a point in the constructed conceptual space. In particular, every ontology concept has been associated to two vectors representing respectively the name and the natural language description of the concept itself [25]. Both the vectors have been calculated applying the mapping formula 10 to such name and description.

Such a mapping makes it possible to evaluate the semantic similarity between concepts, finding relations between them according to the semantic content of the involved concepts. These semantic relations exist apart from those formally defined in the ontology. Such a "geometrical ontological representation" confers to the robot an "intuitive association" capability capable of improving the naturalness of the approach with the user.

7 Dialogue Engine and Communicative Skills

The dialogue engine provides a mechanism to interpret natural languages query given by the user; such s mechanism is based on a classical pattern-matching technique but provides the additive capability of interfacing with the ontology; this is done by transforming natural languages requests into symbolic queries, i.e., queries in the ontology language. Such statements are sent to the ontology engine that calculates the related inferences and provides results in a symbolic form. These symbolic responses are subsequently mapped into natural language sentences that are finally shown to the user [1, 27]. Furthermore, the conversational agent uses the semantic space to both retrieve answers that cannot be activated through the pattern-matching rules or the navigation of the Cyc Ontology, in order to improve its effectiveness in information retrieval [4, 25, 26, 28].

Moreover, the robot can exploit the Emotion Detection module to read to provide the information by adding emotional expressiveness, making the interaction with the human user more natural and catchy.

The dialogue module has been realized by using the A.L.I.C.E. (Artificial Linguistic Internet Computer Entity) technology [2]. The interaction is based on algorithms for automatic detection of patterns in the dialogue data, and it is based on a search algorithm that is a special case of backtracking, depth-first search. Speech recognition can be performed by using the Google Speech recognition APIs [14].

The A.L.I.C.E knowledge base is made of question-answer modules, called *categories* and structured with an XML-like language, named AIML (Artificial Intelligence Mark-up Language). The question is described by the *pattern* tag while the corresponding answer is denoted by the *template* tag. Other tags make it possible the management of more complex answers, allowing defining a specific topic in the dialogue, to remember the context of the conversation, to answer with a random sentence belonging to a list, to give conditional answers, to recursively call other categories, etc.

The simplest form of category constituting the ALICE knowledge base is shown below. Every time the user says “Hello”, the system answers with “Hello there! What’s your name?”

```
<category>
     <pattern>HELLO</pattern>
    <template>Hello there! What's your name?</template>
</category>
```

To communicate with the ontological knowledge base, we have used the CYN tool [7], which constitutes a bridge between an AIML interpreter and Cyc with the aim of answering questions of the user by querying Cyc. This is done by transforming natural languages requests into symbolic queries, that is, commands in the ontology language. Such commands are forwarded to the ontology engine that computes the appropriate inferences and gives results in a symbolic form. The symbolic responses are in turn mapped by the NLP module into natural language sentences that are finally shown to the user through user interfaces [4, 15, 16, 26].

In CYN new tags have been created in order to interact with the Cyc KB. For example, “<cycterm>” translates an English word/phrase into a Cyc symbol; “<cycsystem>” executes a CycL statement and returns the result; “<cycrandom>” executes a CycL query and returns one response at random, etc.

A very simple AIML category concerning the interaction with Cyc is reposted in the following code:

```
<category>
       <pattern>
            WHAT IS A  ROBOT
    </pattern>
        <template>
            A robot is a
                <cycsystem>(cyc-query '(#$isa #$Robot ?X))
                </cycsystem>.
        </template>
 </category>
```

Every time the user says “What is a robot?”, the ResearchCyc is queried by using the construct “(#$isa #$Robot ?X)” and the answer is wrapped as a natural language sentence.

The mechanism which emulates the intuitive/associative capabilities of the robot returns those documents or ontological concepts semantically nearest to a given query. The query is initially mapped into the conceptual space. The LSA-based retrieval mechanism is responsible for the vector distance computation between the vector identifying the query and those associated with either documents or ontological concepts, depending on the choice taken by the user. Such vector similarity is a measure of how semantically related are the entities involved (query and docs/concepts), and it is computed as the cosine between the vectors representing the query and the description of the concepts.

7.1 Expressiveness of the Robot

The robot can add expressiveness while it communicates with the museum visitors explaining details about the artworks.

A greater expressiveness can be obtained by adapting its behavior to the semantic content of what it is saying. For example, by performing gestures, supporting the exposition and expressing emotions that characterize the artworks descriptions. The text conveying the information to provide is analyzed, in the first phase, by the robot. It is equipped with a set of gestures associated to actions or concepts.

The text is segmented in sentences, then, the lemma of verbs and the stem of nouns are detected, and finally they are compared with the annotations of the gestures according to a semantic similarity by considering the shortest path linking the sense of the annotating gesture and the sense of the detected word in the WordNet "is-a" taxonomy [19]. If the semantic similarity is above a given threshold $T_{sim} \in [0, 1]$, whose value is experimentally determined, an action corresponding to the annotated gesture is executed.

The emotion detection module, presented in Sect. 5, is used to detect an emotion related to the text that is being expressed. A specific expressive behavior is associated with the six Ekman basic emotions. Each behavior is obtained by acting on the robot's communicative channels that can be easily correlated by a human observer to the emotions: the LEDs' color, speech parameters such as speed, pitch tone or volume, and the head inclination [5, 12, 17]. These choices are supported by a previous evaluation [31] of the expressiveness of a Pepper robot obtained adapting the robot's gestures, voice intonation, and eyes color according to the emotional content of the text.

8 Implementation Details

The proposed architecture was implemented by exploiting the Pepper humanoid as robotic platform [24], and by using the Robot Operating System (ROS). We created the main modules for accessing the sensors of the robot by defining proper ROS nodes. The speech of the user is acquired by a Bluetooth microphone in order to isolate the voice of the user interacting with the robot, reducing in this manner noise issues. The speech is analyzed by exploiting the Google Speech API, and the corresponding text is analyzed by the retrieval service, as described in the previous sections. Of course, during the interaction, it is important to manage the turns of conversations to avoid recording also the answers of the robot, and it is also important to manage small silences during the emission of an utterance. For this reason, we implemented a finite state machine (FSM) with the "pyfsm" library to isolate the user utterances to send to the Google Speech service. According to the defined FSM, the robot is in a listening state, and when the user starts talking and the audio signal is above a given threshold, it goes into a recording state. If some silence occurs but a timeout is

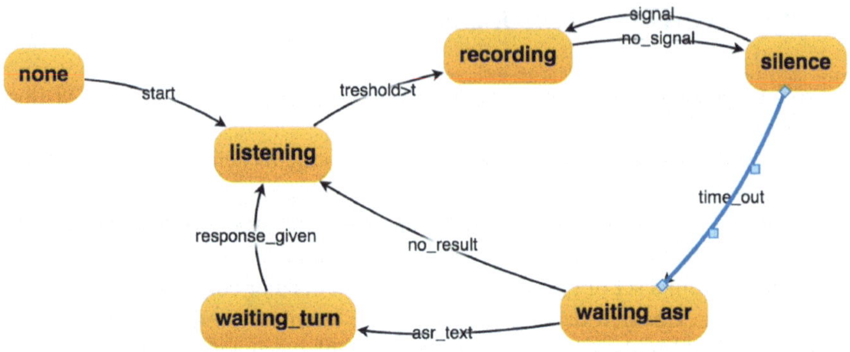

Fig. 3 Finite State Machine for the management of the conversation turns

not reached, the robot remains in the recording state, while after a time out it sends the recorded signal to the Google Speech Recognition service waiting for an answer. The answer is then sent to the retrieval module and the FSM goes in a waiting turn. After that the answer of the robot is given to the user, the FSM goes again in the listening state. This process is illustrated in Fig. 3.

9 Domain Application

In what follows we illustrate a sample of verbal interaction between the robot and the user, showing its retrieval and reasoning capabilities, outlining, in the end, the emotional manifestation and the gestures that it is capable of doing while it is narrating, in order to make the interaction more engaging and amusing for the user.

```
User : Robot, please give me the description of "The Gust" Robot:
This large, 70-gun British warship is
       in distress. A fierce gust\index{Fierce gust} of wind has broken one of
       its masts, and a sail has come loose. In 1672, Willem
       van de Velde II, together with his father, who was
       also a marine painter, entered the service of the
       English court. He made this painting and its
       companion, The Cannon Shot, in England.
User : Does the Gust belong to the seventeenth-century
       period?
Robot: Yes, it is
User : What is the "Battle of Livorno"?
Robot: It is a Painting, which in turn is an ArtObject
User : Is George Hendrik Breitner an artist?
Robot: George Hendrik Breitner is not present in my kb.
       I am giving you the results of wikipedia.
       George Hendrik Breitner (12 September 1857 -
       5 June 1923) was a Dutch painter and photographer.
       An important figure in Amsterdam Impressionism, he
       is noted especially for his paintings of street
```

```
       scenes and harbours in a realistic style. He
       painted en plein air, and became interested in
       photography as a means of documenting street life
       and atmospheric effects rainy weather in particular
       as reference materials for his paintings. Do you
       want to add George Hendrik Breitner in my Cyc
       knowledge Base?
User : Yes please.
Robot: Ok. Added. I will ask to my knowledge engineers to
       insert the relations to complete the insertion.
       I found information about his life on
       wikipedia, I can narrate it to you, if you like.
User : Yes, please.
Robot: George Hendrik Breitner was born in
       Rotterdam, Netherlands...
```

While the narration of the description of "The Gust" is going on, the sentence "A fierce gust of wind has broken one of its masts, and a sail has come loose." is associated to the "sadness" emotion and the robot acts consequently, while the robot is mimicking the "wind" gesture.

10 Conclusions and Future Works

We have illustrated a robotic system capable of interacting with human users in order to inform them about issues regarding the cultural heritage domain with an engaging and amusing manner.

The robotic system is the evolution of previous works developed in the last years, and it takes advantage of a physical humanoid robot, which can lead to better interact with humans. The system is capable of interacting in natural language, to explore a knowledge base both in a traditional, rule-based, manner as well as making sub-symbolic associations between the user queries and concepts stored in the system. Furthermore, when the required information is not available, the robot tries to find it on the Internet.

Moreover, in order to obtain a charming illustration of the information, the robot is capable of showing both emotions and proper gestures while narrating and explaining facts and details to the users.

There are many limitations of the system; the main of them are: the management of the engagement phase with the user, the recognition of the user speech in a noisy and crowded environment, and the robot navigation in the environment.

One of the challenges that we will issue, will be making the interaction more personalized for the kind of people that are interacting with the robot (e.g., young people, adults and elderly).

An evaluation of a prototype of the system will be conducted by involving people who will be asked to fill a satisfaction form after the interaction with the system, in order to identify strengths and weaknesses of the system.

We will adapt and improve previous methodologies to enhance the navigation issues [39], the engagement phase [8], and the user profiling by using mining on social media [35, 36], and by using also more recent knowledge management and deep learning techniques.

Acknowledgements This research has been partially supported by AMICO Project, "Assistenza Medicale In COntextual awareness" decreto di concessione del 10 luglio 2018 prot. n.11598, CUP B46G18000390005; cod ARS01_00900.

References

1. F. Agostaro, A. Augello, G. Pilato, G. Vassallo, S. Gaglio, A conversational agent based on a conceptual interpretation of a data driven semantic space, in AI*IA 2005: Advances in artificial intelligence. Lecture Notes in Computer Sciecnce, (Springer, 2005) **3673**, pp. 381–392
2. Alice, Artificial linguistic computer entity. https://alice.sunlitsurf.com/alice/about.html
3. M. Álvarez, R. Galán, F. Matía, D. Rodríguez-Losada, A. Jiménez, An emotional model for a guide robot. IEEE Trans. Syst. Man Cybern. A Syst. Hum. **40**(5), 982–992 (2010)
4. A. Augello, M. Scriminaci, S. Gaglio, G. Pilato, A modular framework for versatile conversational agent building, in *Complex, Intelligent and Software Intensive Systems (CISIS-2011), International Conference on* (IEEE, 2011), pp. 577–582
5. T. Bänziger, K.R. Scherer, The role of intonation in emotional expressions. Speech Commun. **46**(3–4), 252–267 (2005)
6. A. Chella, R.E. Barone, G. Pilato, R. Sorbello, An emotional storyteller robot, in *AAAI Spring Symposium: Emotion, Personality, and Social Behavior* (2008), pp. 17–22
7. K. Coursey, Living in Cyn: mating AIML and Cyc together with program n (2004). Resources available at: http://www.daxtron.com
8. A. Cuzzocrea, G. Pilato, Taxonomy-based detection of user emotions for advanced artificial intelligent applications. Hybrid Artif. Intell. Syst. Lecture Notes in Computer Science, (Springer, 2018) **10870**, 573–585 (2018)
9. R. Dieter Schraft, B. Graf, A. Traub, D. John, A mobile robot platform for assistance and entertainment. Ind. Robot Int. J. **28**(1), 29–35 (2001)
10. Droids company, (2015). Visited on 9 Nov 2018, http://www.droids-company.com/droidlomnynorio/index.html
11. J. Falconer, Honda's ASIMO get a new job at museum. IEEE Spect. **9**, (2013)
12. J. Feldmaier, T. Marmat, J. Kuhn, K. Diepold, Evaluation of a RGB-LED-based emotion display for affective agents (2016). arXiv preprint arXiv:1612.07303
13. C. Germak, M.L. Lupetti, L. Giuliano, M.E.K. Ng, Robots and cultural heritage: new museum experiences. J. Sci. Technol. Arts **7**(2), 47–57 (2015)
14. Google speech to text apis, https://cloud.google.com/speech-to-tex
15. I. Infantino, G. Pilato, R. Rizzo, F. Vella, A practical approach to humanoid introspection, in *ECAL 2013: The Twelfth European Conference on Artificial Life* (2013), pp. 1005–1006
16. I. Infantino, G. Pilato, R. Rizzo, F. Vella, Humanoid introspection: a practical approach. Int. J. Adv. Robot. Syst. **10**(5), (2013). https://doi.org/10.5772/55919
17. D.O. Johnson, R.H. Cuijpers, D. van der Pol, Imitating human emotions with artificial facial expressions. Int. J. Soc. Robot. **5**(4), 503–513 (2013)
18. T.K. Landauer, P.W. Foltz, D. Laham, An introduction to latent semantic analysis. Discourse Process. **25**(2–3), 259–284 (1998)
19. M. Lingling, H. Runqing, G. Junzhong, A review of semantic similarity measures in wordnet. Int. J. Hybrid Inf. Technol. **6**(1), (2013)
20. I. Macaluso, E. Ardizzone, A. Chella, M. Cossentino, A. Gentile, R. Gradino, I. Infantino, M. Liotta, R. Rizzo, G. Scardino, Experiences with CiceRobot, a museum guide cognitive robot, in AI*IA 2005: Advances in Artificial Intelligence. Lecture Notes in Computer Science, (Springer, 2005) **3673**, pp. 474–482

21. Norio, Un robot pour entrer dans l'histoire (in French). Chateau d'Orion, National Centre for Monuments, France (2015). https://www.monumentsnationaux.fr/Actualites/Un-robot-de-visite-a-Oiron-! Visited on 9 Nov 2018
22. I.R. Nourbakhsh, J. Bobenage, S. Grange, R. Lutz, R. Meyer, A. Soto, An affective mobile robot educator with a full-time job. Artif. Intell. **114**(1–2), 95–124 (1999)
23. Opencyc, http://www.opencyc.org
24. Pepper the humanoid robot, (2015). https://www.softbankrobotics.com/emea/en/pepper Visited on 2 Nov 2018
25. G. Pilato, A. Augello, M. Scriminaci, G. Vassallo, S. Gaglio, Sub-symbolic mapping of Cyc microtheories in data-driven "Conceptual" spaces, in *International Conference on Knowledge-Based and Intelligent Information and Engineering Systems, Lecture Notes in Computer Science* (Springer, Berlin, Heidelberg, 2007), **4692** pp. 156–163
26. G. Pilato, A. Augello, G. Trecarichi, G. Vassallo, S. Gaglio, LSA-enhanced ontologies for information exploration system on cultural heritage, in *AI*IA Workshop for Cultural Heritage—9th Congress of the Italian Association for Artificial Intelligence*, University of Milan Bicocca, 20 Sept 2005
27. G. Pilato, A. Augello, G. Vassallo, Sub-symbolic knowledge representation for evocative chat-bots, in *Interdisciplinary Aspects of Information Systems Studies* (Physica-Verlag Heidelberg, 2008), pp. 343–349
28. G. Pilato, A. Augello, G. Vassallo, S. Gaglio, Sub-symbolic semantic layer in Cyc for intuitive chat-bots, in *International Conference on Semantic Computing (ICSC 2007)* (IEEE, 2007), pp. 121–128
29. G. Pilato, E. D'Avanzo, Data-driven social mood analysis through the conceptualization of emotional fingerprints. Procedia Comput. Sci. **123**, 360–365 (2018)
30. G. Pilato, G. Vassallo, TSVD as a statistical estimator in the latent semantic analysis paradigm. IEEE Trans. Emerg. Top. Comput. **3**(2), 185–192 (2015)
31. I. Rodriguez, A. Manfré, F. Vella, I. Infantino, E. Lazkano, Talking with sentiment: adaptive expression generation behavior for social robots, in *The 19th International Workshop of Physical Agents (WAF)* (2018)
32. O. Stock, E. Not, M. Zancanaro, Intelligent interactive information presentation for cultural tourism, in Multimodal Intelligent Information Presentation (Springer, 2005), **27** pp. 95–111
33. O. Stock, Language-based interfaces and their application for cultural tourism. AI Mag. **22**(1), 85 (2001)
34. Telepresence robot for museums. Commonwealth Scientific and Industrial Research Organisation, Australia (2015)
35. D. Terrana, A. Augello, G. Pilato, Automatic unsupervised polarity detection on a Twitter data stream, in *2014 IEEE International Conference on Semantic Computing* (IEEE, 2014), pp. 128–134
36. D. Terrana, A. Augello, G. Pilato, Facebook users relationships analysis based on sentiment classification, in *2014 IEEE International Conference on Semantic Computing* (IEEE, 2014), pp. 290–296
37. S. Thrun, M. Bennewitz, W. Burgard, A.B. Cremers, F. Dellaert, D. Fox, D.Hähnel, C. Rosenberg, N. Roy, J. Schulte et al., Minerva: A tour-guide robot that learns, in *Annual Conference on Artificial Intelligence* (Springer, 1999), pp. 14–26
38. P.A. Tresset, F. Leymarie, Sketches by Paul the robot, in *Proceedings of the Eighth Annual Symposium on Computational Aesthetics in Graphics, Visualization, and Imaging* (Eurographics Association, 2012), pp. 17–24
39. S. Vitabile, G. Pilato, F. Pullara, F. Sorbello, A navigation system for vision-guided mobile robots, in *Proceedings 10th International Conference on Image Analysis and Processing* (IEEE, 1999), pp. 566–571
40. T. Willeke, C. Kunz, I.R. Nourbakhsh, The history of the Mobot museum robot series: an evolutionary study, in *FLAIRS Conference* (2001), pp. 514–518

Gamification

Playing a game is the voluntary attempt to overcome unnecessary obstacles.
Jane McGonigal

Using Technology and Gamification as a Means of Enhancing Users' Experience at Cultural Heritage Sites

Mark Bugeja and Elaine Marie Grech

Abstract Technology is an application of science that aims to develop and create innovations that improve human society. This process has been ongoing for centuries ever since human progression started with the invention of the first tools. In today's world, with the advanced pace of technological innovations, we may not be taking advantage of all the usefulness that these innovations can offer. In the digital age of big data, information is always present at the user's fingertips. How can old practices and traditional dissemination of information, such as, that used by most museums and cultural heritage sites keep up? In this chapter, we explore several emerging technologies, such as AI, AR, VR and mobile technologies as well as the use of gamification that can serve as aids in presenting cultural heritage information to younger generations in ways that are more appealing to them. Combining gamification with emerging technologies can represent an effective way of increasing visitors' interest and engagement at cultural heritage sites by providing a user-centred experience that is more playful and enjoyable.

1 Introduction

Humans are always evolving, changing and growing. It is this incredible capacity to change that makes humans' potential limitless. Unfortunately, even though our society is mostly based on embracing this rapid growth, certain aspects of society are set in "tradition" or what we sometimes politely refer to as "old fashioned" Presentation of information is a task that most of us do every day. For example, by turning on our smartphones and opening a web browser, we can access news far and wide across the world. We now enjoy things like television on demand, and we do not have to wait to watch our favourite programs. We can subscribe to different services and watch

M. Bugeja (✉) · E. M. Grech
University of Malta, Msida, Malta
e-mail: mark.bugeja@um.edu.mt

E. M. Grech
e-mail: elaine.m.grech@um.edu.mt

D. Seychell and A. Dingli (eds.), *Rediscovering Heritage Through Technology*, Studies in Computational Intelligence 859,
https://doi.org/10.1007/978-3-030-36107-5_4

what we want when we want. Given such personal and fast access to information, are cultural and heritage sites keeping up with today's world? Is placing several historical artefacts or artwork around a room enough? No matter how compelling the subject matter can be, it is the museum designers and curators' responsibility to create an engaging experience for the visitors. Frankly, museums and heritage sites can be thought of as a form of entertainment; a means to learn and understand new things or appreciate the incredible talent of different prominent individuals that emerged throughout history [53]. Just like any other entertainment business, it needs to compete with all other services that offer information on demand. There is already a lot of research that is analysing what visitor expectations are at museum and heritage sites [5, 19]. Other research is also focusing on how younger generations perceive these locations [31]. From work done by these researchers, we can already observe a trend of an emerging generation in society that is looking for more interactive and immersive experiences. In this chapter, a case is made for integrating technology inside museums and heritage sites. Technology is an excellent way of bridging the gap between younger generations and information dissemination [18]. Nonetheless, adding technology for technology's sake seldom bodes well. Gamification focuses on creating a pleasurable and engaging experience by introducing elements of game design in non-game contexts to improve user experience and user engagement [16]. Hughes et al. describe younger generations of museum-goers as yearning for new ways to learn and make museums exciting [31]. In this chapter, we initially set the groundwork needed to explore how gamification can be used to enhance the museum visitors' experience. Then, we examine a series of technological innovations, including mobile technology, Artificial Intelligence (AI), Virtual Reality (VR) and Augmented Reality (AR). We explore how different case studies integrate gamification and technology to innovate and enhance the experience at cultural heritage sites.

2 Gamification

Every organisation that provides a service to customers aims to create an engaging and compelling customer experience for its users. A new business concept that has the potential to increase customers and employees' engagement is gamification. Gamification refers to the process of enhancing a service or an activity by introducing game-thinking and concepts of game design to create a 'gameful' experience to influence people's psychological state and behaviour [33].

Despite being a relatively recent trend in business, gamification has been applied in several contexts, including health and fitness (Fitocracy, Nike+, Zombies Run!), education and learning (Duolingo, DragonBox), book launches (Jay-Z's Decoded), sustainable consumption and recycling (RecycleBank), as well as in cultural heritage contexts (Dallas Museum of Art, OceanGame, Intrigue at the Museum). Gamification can be a valuable resource to enhance visitors' experience at cultural heritage sites by creating an immersive experience that is more interactive, visual and appealing.

Fig. 1 Gamification design framework by Werbach and Hunter 2012

Visitors are allowed to explore cultural sites, such as museums, castles, art galleries, and exhibitions through location-based games, gamified tour guides, quests, challenges, storytelling, and treasure hunts amongst other ways. The skillful design and implementation of game design techniques could provide a means for cultural heritage sites to increase visitors' interest and engagement, while also being a means of achieving a competitive advantage when compared to other non-gamified cultural attractions.

Gamification could result in several positive outcomes, including increased motivation, changes in users' attitudes, increased engagement and enjoyment, increased user activity, increased knowledge, effective learning, and intentions to recommend to others [26, 28]. Technological advancements such as social media platforms, real-time data analytics, cloud services and increased usage of smartphones accelerate and improve the outcomes of gamified initiatives [45]. The adoption of gamification and interactive technologies can contribute to a positive, meaningful experience, through which visitors increase their cultural knowledge in an enjoyable manner.

2.1 Design Framework

One of the most renowned gamification design frameworks is the six-step framework presented by Werbach and Hunter [57] (Fig. 1).

Primarily, organisations would identify the business objectives that they would like to achieve through gamification, that is the goals that the business wants to achieve through gamification. Gamification is not an end in itself, but rather a process that makes consumers' engagement with an offering more enjoyable through the use of game design elements. While not altering the core function of cultural heritage sites, gamification can be used to enhance the experience of the visitors at any museum. Many gamification attempts fail to meet their business objectives due to poor design. Effective gamification is not a simple process of adding points, badges and some other game elements to an existing activity, but a well-thought-out process that is designed with an achievable purpose in mind, such as increasing the number of visitors at a cultural site, increase visitors' engagement or create more awareness and knowledge about cultural heritage sites.

In attaining successful gamification, organisations then define the target behaviours expected from the target audience. The target behaviours encouraged should be consistent with the ultimate business objectives previously defined [57]. Desirable target behaviours may include lengthening the time visitors spend at the museum, encouraging visitors to share their experiences about their visit on social media platforms such as Facebook and Twitter, or attracting more visitors to a cultural heritage site. At this stage, organisations define how the target behaviours will be measured in terms of success metrics or key performance indicators, such as the number of reviews shared online, daily or the monthly number of visits at the cultural heritage site.

A user-centred design process promises the achievement of any defined objectives. User-centred implies that before attempting to gamify the user experience at a cultural heritage site, organisations need to understand the users' profile and their needs, and then seamlessly blend game mechanics and the use of technology to create an engaging and fun experience for the visitors [61]. Gamification is a process that centres around invoking the intrinsic motivation of the users through 'gameful' experiences [32]. As Conaway et al. [11] suggest the customer experience in its entirety is integrated into the process of gamification.

An effective user-centred design relies on a deep understanding of the players' profile. Designers of gamified experiences need to understand users' demographics, such as gender, age group, racial diversity, lifestyles, personalities, motivations, player types, goals and aspirations, as well as previous gaming experience [17, 57]. Since contexts and target users differ, there is no unique gamification formula that fits all purposes.

The next step is to devise the appropriate activity cycles to encourage sustained user engagement [57]. Perceived challenge and users' skills influence the users' engagement and immersion in an activity [13]. Game designers aim to match the perceived level of challenge of activity with the level of skill that players have in performing that task to reach the level of a flow state. A state of flow is an optimal experience characterised by focused concentration, complete immersion and engagement in an activity that is intrinsically motivating for the user [13]. If a task is too challenging for one's skill, the user will get anxious, while if a task is too easy for one's skill, the player will get bored. A sense of mastery and competence follows once players complete the challenge at hand, elevating the players' intrinsic motivation. Immediate and visible feedback needs to be provided following the users' actions, to reinforce and encourage the desired behaviour. Subsequently, a higher level of challenge is needed to match the users' improved skills [13, 29, 57]. A successful gamification design would incorporate evolving levels of challenge in line with the player's skills, motivations and personalities.

An essential element at the core of a successful gamified experience is the notion of fun. Effective gamification allows both the users and the organisation to reach their goals through an experience that is fun for the target audience. While keeping in mind the business objectives, the users' needs, motivations and personalities, designers aim to deploy the appropriate game elements that would be effective to generate the appropriate user engagement and elicit the desired behaviour among players [27, 47]. Some game elements may be more effective than others at generating the desired users' behaviours depending on the players' types, personalities and motivations. Points, levels, badges and achievements may be more useful for players who are after achievement and status. On the other hand, opportunities to collaborate, team competitions, and sharing of gifts to new users may be more productive with socialisers. Gamifying an activity is an iterative and experimental process [57]. Proposed gamified solutions need to be tested with the target audience and further developed to respond to the feedback users may share at a later stage. Re-designing or changing some of the game elements to fine-tune the gamified experience is not unusual. Organisations may assess gamified experiences and gather feedback from users in different ways. Unobtrusive observation is one way of monitoring users' behaviours, conversations and collaboration amongst their peers, during the time users, experience the gamified activity. Other users' feedback methods include qualitative interviews and quantitative surveys that help assess the users' experiences and engagement levels after their use of the gamified activity. Werbach proposes that the gamified experience should be intrinsically motivating and rewarding for the users [56]. After reviewing a variety of museum online educational games, Din et al. [17] identified the core components to design a successful museum game. These principles could also be useful when designing a gamified experience at other cultural sites. Din et al. [17] suggests that (i) the design of the game should create an immersive environment; (ii) players should be given the autonomy to control the flow of the game and choose the forthcoming tasks; (iii) the game should be inviting and challenging for the players with clear goals to be achieved; (iv) the game should allow players to track and manage their progress, and (v) users should fulfil a personal need through the game.

2.2 An Intrinsically Motivating Visitor Experience

People are usually motivated to carry out a task as a result of concurrent extrinsic and intrinsic motivations [13]. Gamification is a tool that facilitates both extrinsic and intrinsic motivation to attain specific goals [50]. The use of rewards, progress, challenge, points, leaderboards, and badges amongst other game elements are external stimuli that motivate individuals extrinsically when they engage in a gamified activity. Such external stimuli are designed to provoke the users' intrinsic motivational needs and have an impact on the users' psychological states [33], encouraging and reinforcing the desired behaviours from the user.

Any given stimulus can be extrinsically and intrinsically motivating at the same time [15]. Rewards such as gaining a badge or an esteemed status may be both intrinsically and extrinsically motivating for some people [15]. However, the use of excessive external motivators may produce a negative effect on the users' intrinsic motivation because the user may focus more on achieving the external rewards and may be less engaged when the external reward is no longer forthcoming [15]. A user-centred framework for meaningful gamification should be designed on intrinsic or internal motivations, rather than extrinsic or external motivations [42].

To make sure that the effect of gamification is not short-term, organisations should aim to create gamified solutions that give users the experience of competence, autonomy and social relatedness [43]. These elements are the three constructs of one of the main theories of intrinsic motivation, the Self-Determination Theory. Ryan et al. [48] suggest that intrinsic motivation is enhanced when the psychological and social needs like competence, autonomy and relatedness are satisfied. In a gamified context, competence is the users' sense of ability and achievement. Users' intrinsic motivation is enhanced when they complete a challenge or a task and experience a feeling of competence and mastery. Autonomy is the users' perceived feeling of control within the gamified experience. Designers of gamified solutions should aim to allow the users to choose the challenge or action that they would like to do to progress to the next phase, rather than controlling the action users need to complete. This would result in a more personalised experience for users. Relatedness is experienced when users feel that they are connected to others in the gamified experience. One possibility to increase social relatedness is to include an option for the users to share their progress through social media platforms.

Behaviour change is at the core of effective gamification being driven by intrinsic and extrinsic motivations [34, 46]. A well-gamified experience that includes the appropriate mix of reinforcements (intrinsic and extrinsic) can lead to desired behavioural changes in business contexts [46], through rewarding the desired users' behaviour. If the desired behavioural outcomes are repeatedly reinforced, individuals participating in the gamified experience will act habitually and routinely, thus relying less on the individuals' cognitive resources [20, 46]. The desired behaviour will become habitual through the cues and rewards that are provided by the gamified experience [20].

3 Technology and Gamification

As technology advances, society is continuously finding new ways of improving how technology is used. From a usability perspective, the more useful a technology is perceived, the more viral the use of the technology becomes. Up to a couple of years ago, technology such as virtual reality (VR) and augmented reality (AR) was unheard of, apart from niche researchers and enthusiasts that had access to it. As these technologies became more affordable and accessible new instances and applications have propelled forward development and research in these areas. A more engaging approach to enhancing usefulness and engagement is by combining technology with gamification. Digital tourism uses different technologies to bring the subject to the user or enhance the site itself [55]. Through the use of 3G and GPS, users can get data on their destination and provide feedback accordingly. Furthermore, through the use of more advanced technologies such as VR and AR tourists are immersed in a virtual world that can give them an exploratory environment which is faithful to their respective touristic destination [25]. The most substantial proportion of expenditure and hassle is often travelling from, and to, a destination. Users tend to, as much as possible, get information on their target destination. Digital tourism is the concept of offering support digitally to these travelling users. There are already various implementations, such as TripAdvisor, Expedia, Tripit, and other travel management systems. The aim is always to recommend and facilitate a users trip. A concept that evolves from digital tourism is that of digitally enhanced tourism [4]. By using the latest technology, tools can be developed that have an intensive effect on a tourist's experience. Places which before could not be reached can now become accessible through applications which make use of VR or AR [49]. Depending on the destination and context, user experience designer and researchers can use different techniques to accommodate the user. These tools include interactive maps and virtual guides for museums and heritage sites. In order to design and develop applications which make use of VR and AR technology [25], there is an essential factor that needs to be taken into account, that is, that, no assumptions should be made on how well versed users are in using computing technology. Most well-crafted applications today are designed using user experience and user-centred design principles [8], which enable the users to immerse themselves in an experience and, by sharing with others, become, ambassadors of the technology themselves [4]. The following section presents an overview of different technologies as well as use cases of applications and systems that combine technology with gamification. Not all presented use cases were applied to heritage. Nonetheless, an argument is presented on why the approach is easily adaptable for heritage use.

3.1 Mixed Reality

Computing technology has also enabled us to bring to the consumer devices, which were previously thought to be difficult or impossible to produce. One such technology is VR, the result of years of human progression which ties down the mental immersion with the physical. VR can be defined as a physical and mental immersion into a virtual world. The key concept is the immersion aspect, which refers to an emotional or mental state—a feeling of being involved in the experience [55]. Through a VR headset, a user can enter a virtual world physically through visual and multi-sensory feedback and mentally through interactive elements within the virtual world [25]. The user's brain convolves the amalgamation of sensory data, and these processes mentally puts the user in the virtual world. Today, virtual reality is being used in a variety of applications such as educational aids, visualisation of volumetric data, video games and in tourism, museums and heritage [8]. Statistics show that the tourism sector has been increasing rapidly over the last decade [4]. Not much has been done to address heritage touristic sites which are inaccessible. Such sites are very difficult or impossible to reach or are in such a bad state that an increase in tourists might harm the surrounding area [49]. Through VR, this issue can be addressed by developing photorealistic virtual environments or scanned models of heritage sites.

Users are then able to explore these areas at their leisure [55]. This new kind of digital tourism is advantageous as it allows users to enjoy heritage sites even though they might usually be unable to due to some mobility impaired issues or impossible physical access to these sites.

3.1.1 Mixed Reality Use Cases

Gamification pairs very well with VR as the technology helps users immerse themselves inside the virtual environment, thus automatically enhancing engagement. Villagrasa et al. [54] developed a gamified VR application as a teaching aid for university architecture students. Game mechanics such as leaderboards, badges and loyalty programs were used to create a collaborative VR learning environment. The application was developed using the 'Unity3D' game engine, a development tool that the researchers identified, amongst other choices as the most reliable when working with VR. The evaluation of the project measured the students progress in terms of motivation, satisfaction and academic performance.

The students were given a set of challenges and tasks to build 3D architecture models inside a virtual setting that the authors called "the virtual island" as seen in Fig. 2. Each numbered block inside the virtual island is a task that students had to collaborate to complete. The tasks required students to design different architectural models (see Fig. 3) and individually rate the best ones to increase positive competitiveness within the classroom environment. VR, in this case, was used to access the designs of the finished design and as a means to explore the island.

This use case differs from other VR applications used in heritage as with most heritage and museum VR applications the user is going to go through an experience intended for a single use or limited use. On the other hand, the application developed by Villagrossa et al. [54] is aimed for students that get to use the system over a long period. Thus, the inclusion of leaderboards and badges is meaningful, as the users, in this case, the students, is a returning user that is exposed to the VR application throughout the course (one semester). Moreover, the authors describe that their research showed that the students still needed the educators to teach them more advanced concepts and the tools and applications developed do not replace the educator in the classroom. Nevertheless, the use of VR was positively viewed by the students and increased engagement in the subject matter.

DynaMus [36] is a framework built upon the concept of virtual museums. The authors reviewed several virtual museum applications and found that these applications did not contain much content, such as, an adequate amount of artefacts and rooms that make the experience worthwhile for a user. Most of these reviewed museums only feature a small number of artefacts and the information explaining what the objects are is also very poor. The authors also note that their framework is built upon the concept of serious games and gamification and creates an environment for museum designers and content creators to create virtual spaces where they can host their museums. A serious game is different from a gamified application, and these games are complete games developed for education to exploit game mechanics to

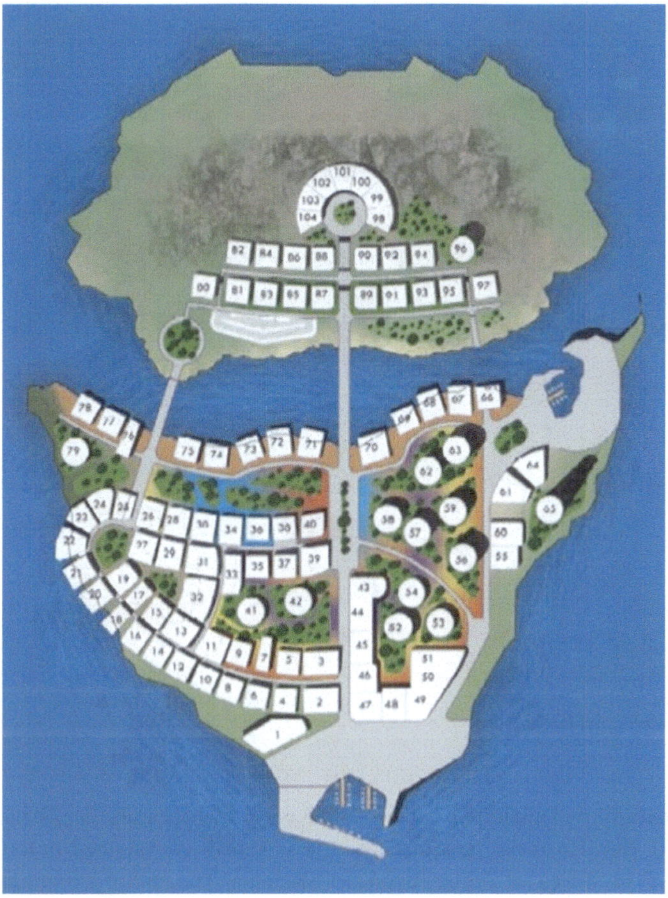

Fig. 2 The virtual island

educate the users on a concept or subject. The DynaMus system does not present itself as a game. Thus, it falls more under the area of gamified applications especially given that the authors state this in their work "that although there is almost no classroom electronic gaming, the prospect of it becoming possible and the adoption of DynaMus is highly anticipated" [36]. This statement implies that the system can be developed into a serious game in the future. DynaMus is a 3D-content management system that uses search engines such as Google web service and Europena web service to gather information about historical artefacts, the content creator has an option to upload his 3D or 2D model to add to the information mined from the search engines (see Fig. 4). There are also various options in the system to let the content creators edit information. The researchers evaluated the framework and the developed system by creating a case study of a virtual museum. The virtual environment called "Synthesis" a virtual museum based upon a collection of artwork mainly

Fig. 3 Example pavillion designs

by the artist Spyros Papaloukas, an esteemed early 20th-century Greek painter (see Fig. 5) [35]. Similar to the previous case the system was evaluated in terms of the "classroom" in this cases young secondary students. The author's findings indicate that students, tutors and even parents were positively pleased with the experience [35]. Students expressed scepticism with regards to tutors being able to adapt to the technology. Although the system has not been tested on VR, the authors describe the framework as VR ready given that it was developed using the "Unity3D" technology that similar to the previous use case, the authors identified as the most reliable for developing applications of this kind. This particular use case is interesting as users do not physically need to be at the museum in order to explore it. Furthermore, the system also allows users to become content creators. Thus, adding a fun gaming element. Content can be downloaded directly through the search engines and the final artefacts placed inside the virtual museum.

Virtual reality creates an immersive almost out of body experience, Augmented Reality (AR) on the other hand, through the use of a mobile phone camera or webcam, overlays 3D models, artwork or effects on the real world. One of the best works to exemplify the use of AR in heritage and culture is a virtual AR system developed by White et al. called ARCO (Augmented Representation of Cultural Objects) [58]. The ARCO system developed in 2004 is a content management system that allows a museum to create a browsable virtual web museum. Using imagery and other distinct

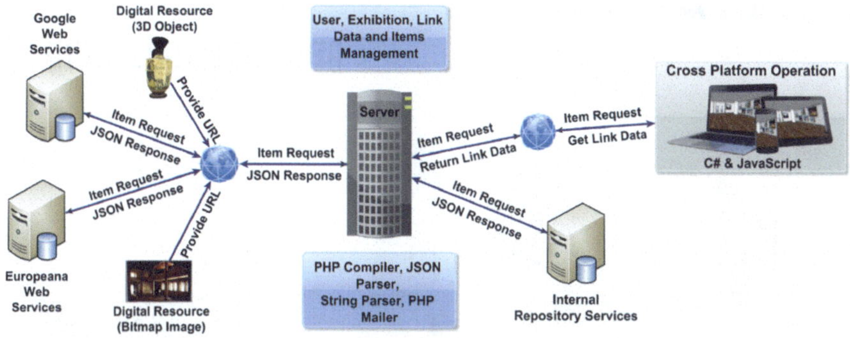

Fig. 4 General technological framework of Dynamus

Fig. 5 Real-time screen space effects

features in the physical museum, a user can also download and use an AR application developed for the ARCO system to bring the museum to life. Through the mobile application, the user can zoom in on different unique sections of the museum, and 3D models are overlayed on the real world on the user's smartphone (see Fig. 6). Although the authors do not specifically mention gamification, the system was built with user engagement in mind. One of the functionalities of the application allows the content creator of the AR museum to build quizzes. Visitors using the application are asked questions related to the AR or physical exhibits, as shown in Fig. 7. This gamification aspect was further highlighted in the work of Sylaiou et al. [51] that

Fig. 6 Example AR scene

have evaluated the system and concluded that the gaming aspect of the system is the most promising. Especially, when engaging with the younger generation, a point that correlated perfectly with the work of Dingli et al. [18] that states that games and technology provide better tools for communicating information to the new wave of emerging adults.

One of the factors that contribute to the increased adoption of and use of AR applications and positively increases engagement is the visual appeal of the 3D models used, and beautiful things work better [44]. Research conducted by Chung et al. [10] analysed an AR application developed for Deoksugung Palace, South Korea (Fig. 8) where a number of aspects where measured in order to asses the technology readiness (TR), the visual impact of the 3D models used in the AR application as well as the situational factor [10]. The application's main role is that of facilitating a user (tourist) journey and acts as a guide of sorts. The study concluded that the TR and visual impact of the application directly influenced the perceived ease of use and usefulness of the AR application. These findings were not surprising as they already correlate with other research in the area [9, 38, 52]. The work also implies that by facilitating the ease of use, the user can be nudged into different locations when visiting a heritage site. Such an application is useful when promoting the conservation of physical sites. Also, combining the technology itself with gamification might improve the results given that user engagement can be heightened considerably if

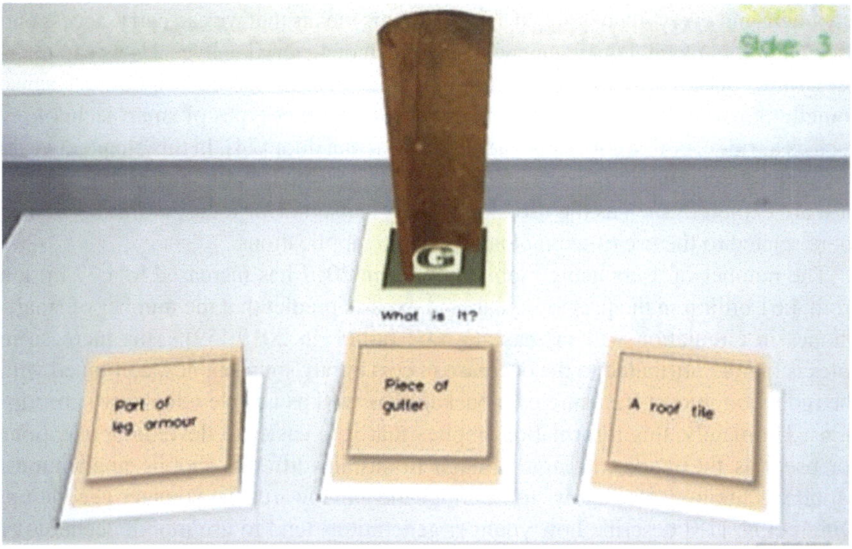

Fig. 7 Example quiz scene

Fig. 8 AR Application in Deoksugung Palace

the right elements are applied. Although the results show positive trends for the adoption of AR applications, the authors note that they conducted the study amongst a sample of 145 users. Thus, the results are not necessarily statistically significant. Furthermore, the results were measured upon one single AR application. Therefore, a wider study sampling different AR applications might yield more conclusive results.

3.2 Mobile Computing

The creation of smartphone technology is a development upon the concept of ubiquitous computing. This concept or rather paradigm defines computing as happening anywhere and everywhere [39]. In a nutshell, this means that we can carry, access and use computing technology at any point in time in our everyday lives. Users can interact with technology by using different machines be it in a more "traditional" sense something like a smartphone or home computer or other types of smart technology such as a fridge, coffee cup, raincoat, toothbrush or a door [24]. In this chapter, we do not analyse technology such as the Internet of Things (IoT)[1] although certain aspects of it are explored, such as the use of proximity sensors. Our focus, in this section, is more related to the use of technology in mobile applications.

The number of users using mobile phones in 2017 has increased to 4.77 billion from 4.61 billion in the previous year, and experts predict that the number of smartphones in circulation will increase to 5.07 billion in 2019 [59]. This increase in sales is mainly attributed to the decrease in cost to buy smartphones, combined with the reduction in cost for using data packages as well as mobile telephony subscriptions. Ultimately, this information implies that it is easier to develop applications for users as the market is large enough to sustain different mobile applications. Mobile technology also allows us to bridge the gap towards the younger generation. Dingli et al. [18] describe how younger generations tend to use mobile technology as a primary means of communications. Thus, smartphones are the best tools to use to communicate information effectively to younger generations.

3.2.1 Mobile Computing Use Cases

There are multiple use cases that developed gamified mobile applications for heritage sites [7, 14, 30]. Some of these applications have already been explored concerning other technologies such as AR and VR. The applications analysed in this section makes use of the native GPS sensor present in most smartphone devices in order to encourage tourist exploration of heritage sites. These applications apply gamification elements as a tool to promote exploration. The research conducted by Bugeja et al. [6] promotes tourist exploration through the concept of "selfie tourism". An initial study established the use of photography and "selfies" as an element that can persuade users to explore different areas while abroad. This research includes the use of an application that contains the best "photographic" sites. The authors then developed a mobile application based in Valletta, the capital city of Malta, that presents the user with a journey through Valletta's most photographic and memorable sites.

In addition, the application was further gamified by rewarding users with different camera filters that are unlocked every time a user visits a new location (Fig. 9). The application also provides information related to the site, such as descriptions and information related to national monuments. Besides, the user can record and map his journey to further reward them for exploring the city (Fig. 10).

The evaluated results were encouraging, but further research needs to be done to consolidate the results further. The final evaluation was conducted by interviewing some participants that used the application and might not necessarily be represen-

[1]Reference [59] Internet of Things.

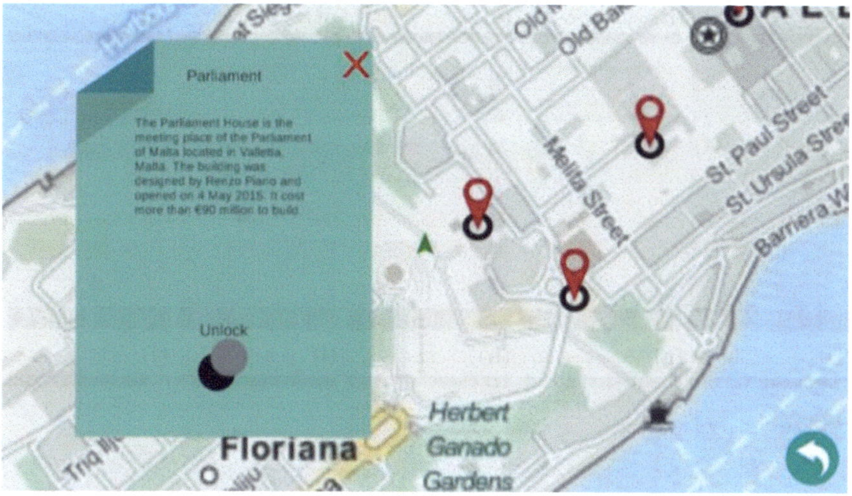

Fig. 9 Application interface when unlocking a heritage site

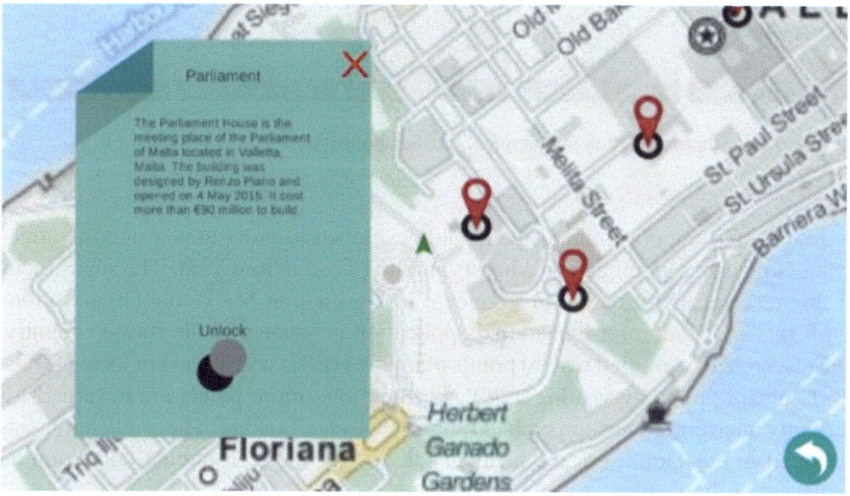

Fig. 10 User journey through pictures

tative of the broader population. The participants indicated that they are now more willing to download similar applications with one particular participant highlighting the fun and playful aspect of the application, "It was fun and interactive; definitely, something I would use again and the information provided turned it into a learning experience. Also, knowing when to take a photo makes it easier to enjoy one's surrounding without having to worry that a photo was not taken in time".

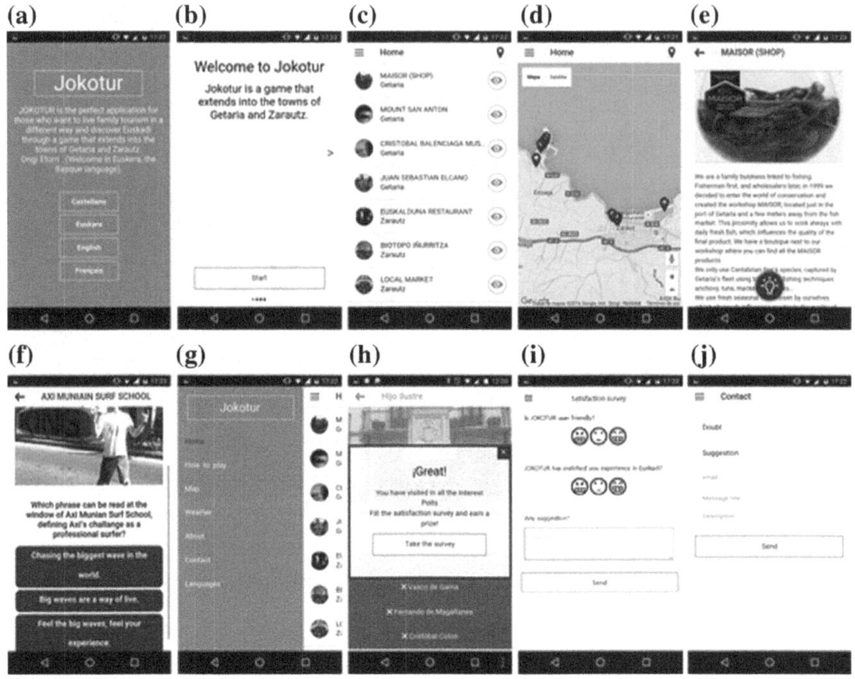

Fig. 11 Screenshots of the Jokotur gamified mobile application

Jokotur is an application designed and developed in order to research the effectiveness of using gamification when promoting destinations [23]. The main focus of this research is to market this concept to Destination Marketing Organisations (DMO). The location for the studied application is situated in the Basque country. The researchers identified several points of interest (POI) with the aid of local DMOs. Users obtain points by visiting a POI. Furthermore, more points are rewarded for correctly answering questions related to the POI (see Fig. 11). The application was designed to put families at the heart of the design, thus, promoting collaboration between family members to gain points.

The tourists are further encouraged to visit the POI as once any of the three out of 8 locations are explored the users will gain enough points to visit a tourist office and gain a real reward. The evaluation of the application is conducted through an in-app questionnaire that assesses the effectiveness and easiness of the gamified mobile experience. Out of the 124 evaluated tourists, the app scored 2.7 out of 3 with 22 of the tourists exploring at least 3 of the POI shown in the application. The authors note that one of the limitations of the system is finding ways to promote the application to tourists [22] correctly. The authors concluded that more research is required in order to find better ways to market similar mobile applications, especially since users are sometimes reluctant to download mobile applications that take up storage space.

3.3 *Artificial Intelligence*

Artificial Intelligence (AI) is the development of technology that can think or act intelligently or in a broader sense like a human. In the past decade, we have seen a rise in research and development in AI largely due to the improvement in computing technology. This surge in computing research resulted in concepts that were discussed but not completely proven decades ago to be implemented and used. Notably, the area of Machine Learning (ML) which is the study of statistical model computationally trained using a dataset in order to classify or predict information. This research area has given rise to Deep Learning (DL) that uses massive amounts of data to train machines on different types of data such as images, text and sound to gain more meaningful prediction [37]. Deep Learning is the state of the art approach to solving machine learning problems, given how effective these models are when successfully applied to various fields of study, such as speech recognition and computer vision. Furthermore, with the advent of big data, companies such as Google, Microsoft and Apple that have access to large volumes of data have consistently pushed the field of Deep Learning and have applied it to well-known systems such as Siri for Apple [21] and Google Translate for Google [60]. Due to the rapid advancement of this technology, researchers have developed technologies that can interact better with users such as chatbots.

3.3.1 Chat Bots

Chatbots systems are simply a dialogue system based on natural language, for this reason, they can be used as interfaces to a vast number of applications including entertaining applications, educational applications, e-learning platforms, research engines and e-commerce website navigation. Several approaches use natural language to model the interfaces or try to combine simple technology of chatbots systems, with more sophisticated techniques [1]. What is lacking in AI, and, as a consequence, in conversational systems is the creative capability of humans that gives humans intuitively the ability to create new original ideas. The CyN project [12] aimed to improve upon the human aspect of chatbots through an experiment that links ALICE a pattern matching interpreter and prominent conversational system, to OpenCyc [40] a large commonsense knowledge base. During the conversation, the dialogue between human beings moves from one topic to another. This communication flow is due to human beings having great associative capabilities. Research has already been done in this area to design a framework for a human-like information supplier. Such a system can combine "intuition" with the more common rules-based knowledge techniques. The work developed by Augello et al. [2] adds the human element by using LSA (Latent Semantic Analysis) methodology: this concept infers the latent relational knowledge between the words belonging to a dataset containing a large collection of documents. Other approaches combine Deep Learning to teach chatbots to converse in a human-like fashion, notably the latest work developed by Google [41].

3.3.2 Chat Bots Use Case

The potential of chatbots combined with gamification and used in a cultural context is limitless. Through the technology, users are engaged further as they can directly talk to an "individual" that might present itself as a guide. Furthermore, some creative games and solutions can be developed using this technology. Most of the problems faced with implementing such a system are how to develop a conversationally perfect solution. Although such a solution might not be very far off. One application that pioneers the usage of chatbots in heritage is the research conducted by Augello et al. [3]. The idea behind the development of the application is to replace audio/visual guides usually present at museums and other heritage sites by developing a pervasive virtual guide. The architecture of the system is designed to use a conversational agent that can answer questions related to exhibits and other objects in the heritage site. Moreover, through the use of RFID information inferences can be made by the system to better answer questions asked by users (Fig. 12). The conversation is built using a knowledge base and ontologies to deconstruct a query into essential parts and construct a message to reply to the user. A prototype system has been developed

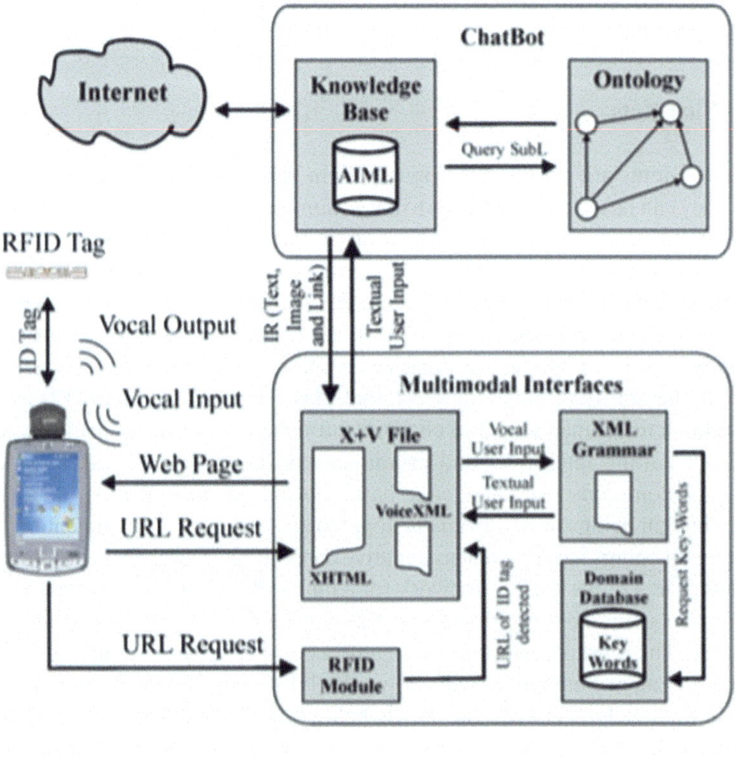

Fig. 12 System architecture

for "Museo Archeologico Della Valle Dei Templi", but no results are presented to showcase user review and rating of the approach.

4 Conclusion

Research has identified a demand from younger generations to create more meaningful experiences in cultural heritage contexts. In this chapter, we explore how cultural heritage sites can use different innovative technologies to enhance the experience for their visitors through the use of mobile technology, Virtual Reality, Augmented Reality, AI agents and location-based technology, and gamification. The case studies explored in this chapter present a combination of different technologies and gamified experiences that motivate and encourage users to explore heritage and museum sites, enhance user engagement and knowledge.

By integrating new technologies with gamified experiences, cultural heritage sites will appeal more to what visitors look for in this digital era. Further research related to the implementation and the effect of these technologies and game elements, such as points, leaderboards, badges, challenges and storytelling/narrative on user experience is required.

References

1. S.A. Abdul-Kader, J. Woods, Survey on chatbot design techniques in speech conversation systems. Int. J. Adv. Comput. Sci. Appl. **6**(7) (2015)
2. A. Augello, G. Pilato, G. Vassallo, S. Gaglio, A semantic layer on semi-structured data sources for intuitive chatbots, in *2009 International Conference on Complex, Intelligent and Software Intensive Systems* (IEEE, 2009), pp. 760–765
3. A. Augello, A. Santangelo, S. Sorce, G. Pilato, A. Gentile, A. Genco, S. Gaglio, A multimodal interaction guide for pervasive services access, in *IEEE International Conference on Pervasive Services* (IEEE, 2007), pp. 250–256
4. D. Benyon, A. Quigley, B. O'Keefe, G. Riva, Presence and digital tourism. AI Soc. **29**(4), 521–529 (2014)
5. J.G. Brida, M. Meleddu, M. Pulina, Understanding museum visitors' experience: a comparative study. J. Cult. Herit. Manag. Sustain. Dev. **6**(1), 47–71 (2016)
6. M. Bugeja, A. Dingli, D. Seychell, Selfie as a motivational tool for city exploration, in *International Association for Development of the Information Society* (2018)
7. A. Bujari, M. Ciman, O. Gaggi, C.E. Palazzi, Using gamification to discover cultural heritage locations from geo-tagged photos. Pers. Ubiquitous Comput. **21**(2), 235–252 (2017)
8. D. Castilla, A. Garcia-Palacios, J. BretóN-LóPez, I. Miralles, R.M. BañOs, E. Etchemendy, L. Farfallini, C. Botella, Process of design and usability evaluation of a telepsychology web and virtual reality system for the elderly: Butler. Int. J. Hum.-Comput. Stud. **71**(3), 350–362 (2013)
9. C.-F. Chen, D. Tsai, How destination image and evaluative factors affect behavioral intentions? Tour. Manag. **28**(4), 1115–1122 (2007)
10. N. Chung, H. Han, Y. Joun, Tourists' intention to visit a destination: the role of augmented reality (ar) application for a heritage site. Comput. Hum. Behav. **50**, 588–599 (2015)
11. R. Conaway, M.C. Garay, Gamification and service marketing. SpringerPlus **3**(1), 653 (2014)

12. K. Coursey, Living in cyn: mating aiml and cyc together with program n (2004), http://www.daxtron.com
13. M. Csikszentmihalyi, *Intrinsic motivation in museums: What makes visitors want to learn* (Public institutions for personal learning, Establishing a research agenda, 1995)
14. S. de los Ríos, M.F. Cabrera-Umpiérrez, M.T. Arredondo, M. Páramo, B. Baranski, J. Meis, M. Gerhard, B. Prados, L. Pérez, M. del Mar Villafranca, Using augmented reality and social media in mobile applications to engage people on cultural sites, in *International Conference on Universal Access in Human-Computer Interaction* (Springer, 2014), pp. 662–672
15. E.L. Deci, R. Koestner, R.M. Ryan, A meta-analytic review of experiments examining the effects of extrinsic rewards on intrinsic motivation. Psychol. Bull. **125**(6), 627 (1999)
16. S. Deterding, R. Khaled, L.E. Nacke, D. Dixon, Gamification: toward a definition, in *CHI 2011 Gamification Workshop Proceedings*, vol. 12 (Vancouver BC, Canada, 2011)
17. H.W.-H. Din, Play to learn: exploring online educational games in museums, in *ACM SIGGRAPH 2006 Educators program* (ACM, 2006), p. 13
18. A. Dingli, D. Seychell, *The New Digital Natives: Cutting the Chord* (Springer, Berlin, 2015)
19. K. Drotner, L.V. Knudsen, C.H. Mortenesen, Young people's own museum views. Mus. Manag. Curatorship **32**(5), 456–472 (2017)
20. C. Duhigg, *The Power of Habit: Why We Do what We Do in Life and Business* (Random House, 2012)
21. A. Gandomi, M. Haider, Beyond the hype: big data concepts, methods, and analytics. Int. J. Inf. Manag. **35**(2), 137–144 (2015)
22. A. Garcia, M.T. Linaza, A. Gutierrez, E. Garcia, Validation of a gamified mobile experience by dmos, in *Information and Communication Technologies in Tourism 2017* (Springer, 2017), pp. 331–343
23. A. Garcia, M.T. Linaza, A. Gutierrez, E. Garcia, Gamified mobile experiences: smart technologies for tourism destinations. Tour. Rev. **74**(1), 30–49 (2019)
24. A. Greenfield, *Everyware: The Dawning Age of Ubiquitous Computing* (New Riders, 2010)
25. J.P. Guerra, M.M. Pinto, C. Beato, Virtual reality-shows a new vision for tourism and heritage. Eur. Sci. J. ESJ **11**(9) (2015)
26. J. Hamari, Transforming homo economicus into homo ludens: a field experiment on gamification in a utilitarian peer-to-peer trading service. Electron. Commer. Res. Appl. **12**(4), 236–245 (2013)
27. J. Hamari, L. Hassan, A. Dias, Gamification, quantified-self or social networking? matching users' goals with motivational technology. User Model. User-Adapt. Interact. **28**(1), 35–74 (2018)
28. J. Hamari, J. Koivisto, H. Sarsa et al., Does gamification work?-a literature review of empirical studies on gamification. HICSS **14**, 3025–3034 (2014)
29. J. Hamari, D.J. Shernoff, E. Rowe, B. Coller, J. Asbell-Clarke, T. Edwards, Challenging games help students learn: an empirical study on engagement, flow and immersion in game-based learning. Comput. Hum. Behav. **54**, 170–179 (2016)
30. R. Hammady, M. Ma, N. Temple, Augmented reality and gamification in heritage museums, in *Joint International Conference on Serious Games* (Springer, 2016), pp. 181–187
31. K. Hughes, G. Moscardo, Connecting with new audiences: exploring the impact of mobile communication devices on the experiences of young adults in museums. Visit. Stud. **20**(1), 33–55 (2017)
32. K. Huotari, J. Hamari, Defining gamification: a service marketing perspective, in *Proceeding of the 16th International Academic MindTrek Conference* (ACM, 2012), pp. 17–22
33. K. Huotari, J. Hamari, A definition for gamification: anchoring gamification in the service marketing literature. Electron. Mark. **27**(1), 21–31 (2017)
34. D.L. Kappen, L.E. Nacke, The kaleidoscope of effective gamification: deconstructing gamification in business applications, in *Proceedings of the First International Conference on Gameful Design, Research, and Applications* (ACM, 2013), pp. 119–122
35. C. Kiourt, A. Koutsoudis, S. Markantonatou, G. Pavlidis, The 'synthesis' virtual museum. Mediterr. Archaeol. Archaeom. **16**(5) (2016)

36. C. Kiourt, A. Koutsoudis, G. Pavlidis, Dynamus: a fully dynamic 3d virtual museum framework. J. Cult. Herit. **22**, 984–991 (2016)
37. Y. LeCun, Y. Bengio, G. Hinton, Deep learning. Nature **521**(7553), 436 (2015)
38. J.-S.C. Lin, H.-C. Chang, The role of technology readiness in self-service technology acceptance. Manag. Serv. Qual. Int. J. **21**(4), 424–444 (2011)
39. K. Lyytinen, Y. Yoo, Ubiquitous computing. Commun. ACM **45**(12), 63–96 (2002)
40. C. Matuszek, M. Witbrock, J. Cabral, J. DeOliveira, An introduction to the syntax and content of cyc (UMBC Computer Science and Electrical Engineering Department Collection, 2006)
41. M. McTear, Conversational modelling for chatbots: current approaches and future directions. Technical report, Ulster University, Ireland (2018)
42. S. Nicholson, Strategies for meaningful gamification: concepts behind transformative play and participatory museums. *Meaningful Play* (2012)
43. S. Nicholson, A recipe for meaningful gamification, in *Gamification in Education and Business* (Springer, 2015), pp. 1–20
44. D. Norman, Emotion & design: attractive things work better. Interactions **9**(4), 36–42 (2002)
45. D. Palmer, S. Lunceford, A.J. Patton, The engagement economy: how gamification is reshaping businesses. Deloitte Rev. **11**, 52–69 (2012)
46. K. Robson, K. Plangger, J.H. Kietzmann, I. McCarthy, L. Pitt, Is it all a game? Understanding the principles of gamification. Bus. Horiz. **58**(4), 411–420 (2015)
47. K. Robson, K. Plangger, J.H. Kietzmann, I. McCarthy, L. Pitt, Game on: engaging customers and employees through gamification. Bus. Horiz. **59**(1), 29–36 (2016)
48. R.M. Ryan, E.L. Deci, Self-determination theory and the facilitation of intrinsic motivation, social development, and well-being. Am. Psychol. **55**(1), 68 (2000)
49. A. Sambhanthan, A. Good, Enhancing tourism destination accessibility in developing countries through virtual worlds (2013), arXiv:1306.1630
50. K. Seaborn, D.I. Fels, Gamification in theory and action: a survey. Int. J. Hum. Comput. Stud. **74**, 14–31 (2015)
51. S. Sylaiou, M. Economou, A. Karoulis, M. White, The evaluation of arco: a lesson in curatorial competence and intuition with new technology. Comput. Entertain. (CIE) **6**(2), 23 (2008)
52. O. Turel, A. Serenko, N. Bontis, User acceptance of hedonic digital artifacts: a theory of consumption values perspective. Inf. Manag. **47**(1), 53–59 (2010)
53. I. Van Aalst, I. Boogaarts, From museum to mass entertainment: the evolution of the role of museums in cities. Eur. Urban Region. Stud. **9**(3), 195–209 (2002)
54. S. Villagrasa, D. Fonseca, J. Durán, Teaching case: applying gamification techniques and virtual reality for learning building engineering 3d arts, in *Proceedings of the Second International Conference on Technological Ecosystems for Enhancing Multiculturality* (ACM, 2014), pp. 171–177
55. C. Wang, Application of virtual reality technology in digital tourism, in *2011 Third International Conference on Multimedia Information Networking and Security* (IEEE, 2011), pp. 537–541
56. K. Werbach, (re) defining gamification: a process approach, in *International Conference on Persuasive Technology* (Springer, 2014), pp. 266–272
57. K. Werbach, D. Hunter, For the Win: How Game Thinking Can Revolutionize Your Business (Wharton Digital Press, 2012)
58. M. White, N. Mourkoussis, J. Darcy, P. Petridis, F. Liarokapis, P. Lister, K. Walczak, R. Wojciechowski, W. Cellary, J. Chmielewski et al., Arco-an architecture for digitization, management and presentation of virtual exhibitions, in *Proceedings Computer Graphics International, 2004* (IEEE, 2004), pp. 622–625
59. F. Wortmann, K. Flüchter, Internet of things. Bus. Inf. Syst. Eng. **57**(3), 221–224 (2015)
60. Y. Wu, M. Schuster, Z. Chen, Q.V. Le, M. Norouzi, W. Macherey, M. Krikun, Y. Cao, Q. Gao, K. Macherey et al., Google's neural machine translation system: bridging the gap between human and machine translation (2016), arXiv:1609.08144
61. F. Xu, F. Tian, D. Buhalis, J. Weber, H. Zhang, Tourists as mobile gamers: gamification for tourism marketing. J. Travel Tour. Mark. **33**(8), 1124–1142 (2016)

Win the City Game

Foaad Haddod and Alexiei Dingli

Abstract Serious games continue to attract the interest of developers and users. This project aims to design a serious game called Win the City to explore famous attractions e.g. buildings in the city of Valletta in Malta. The Adobe PhoneGap, Google Maps API, and Firebase platform for Android were used to design the prototype for the game. The game uses the real Valletta map to locate the targeted buildings as presented on Google maps. All the provided details and options on Google Maps are shown on the game's prototype. The game can be installed on mobile devices that support the Android platform. The designed prototype was tested by a number of users (residents and visitors) to collect feedback about the game idea, its design, the included functionalities and features, and its overall usability. It's expected that this game will enhance the knowledge of local people as well as the visitors about the targeted famous buildings and sites. Also, the game will provide valuable feedback to the city council about how such buildings are located, frequency of visits, and any physical issues with the buildings. On the other hand, playing the game will promote more exercise activities and enhance the wellbeing of the players.

1 Introduction

Serious games usually refer to games used for training, advertising, simulation, or education that is designed to run on personal computers or video game consoles [17]. Serious games field has been growing in a rapid pace and potential uses of related applications have also expanded in different domains, including education, training, well-being, advertisement, cultural heritage, interpersonal communication, and health care [10]. The current wave of 'Serious Games' can be considered to have begun in 2002 and in 2010, the worldwide Serious Games market was worth 1.5

F. Haddod · A. Dingli (✉)
University of Malta, Msida, Malta
e-mail: alexiei.dingli@um.edu.mt

F. Haddod
e-mail: foaad.haddod@um.edu.mt

D. Seychell and A. Dingli (eds.), *Rediscovering Heritage Through Technology*, Studies in Computational Intelligence 859,
https://doi.org/10.1007/978-3-030-36107-5_5

billion [5]. Recent ICT advances have led to the implementation of realistic virtual environments and simulations, where players can live compelling adventures while acquiring, practising and verifying knowledge, according to various pedagogical paradigms [3].

A major challenge lies in translating interest and potential into actual adoption and use. SGs must demonstrate the transfer of learning (to be 'serious'), whilst also remaining engaging and entertaining (to be 'games'). The balance between fun and educational measures should be targeted throughout the development starting from the design phase [20]. Modern theories of effective learning suggest that learning is most effective when it is active, experiential, situated, problem-based and provides immediate feedback [8, 9]. Serious games have been used to change behaviours and attitudes more broadly in the areas of health, public policy and advertising as well as education and training [6].

In this project, the design and development of a serious game (Win the City) to engage and motivate players to explore the city of Valletta in Malta is described. Ways in which intelligence can be added to the game interface to create a more effective user experience is also explored and presented. Win the City game aims to attract citizens, and visitors, of Valetta city to discover its many attractions. Valetta, as the capital of Malta, has a variety of historical buildings; Churches, Monuments, Gardens…etc. Creating a serious game allows users to learn, discover, and visit all the city's attraction sites. Based on the author's knowledge, no similar previous game was designed and thus, Win the City game is a first of its kind and will be a novel work in this field. This serious game was designed as a part of the requirements for an Advanced Intelligent Interfaces graduate course at the University of Malta. It included a prototype of the game, a design of the user interface for the mobile application and the computational model of the game. The article is organized as follows: in Sect. 2 a relevant literature on serious games was reviewed. In Sect. 3, a demonstration of the game design of the digital prototype, the user interface design, and a Firebase project of the serious game titled Win the City. In Sect. 4, the game experiment and testing are presented and the suggested questions to be asked to the residents and experts are provided. In Sect. 5, provides an overview of the results and evaluations. Conclusions and further research in this area are presented in Sect. 6.

2 Literature Review

2.1 Origin of Serious Games

Over the last 40 years, computer games have increasingly replaced more traditional games as leisure activities and have had a transformational impact on how we spend our leisure time [8]. The idea of using games for purposes other than fun was first formulated in the book Serious Games by Clark C. Abt (1975) [7]. The term serious game as it applies to digital games was coined by Ben Sawyer in a paper on the potential of using digital games for policy making [23]. The first appearance of serious games is believed to have been in the U.S., in 2002, when America's Army game was developed and distributed, free-of-charge, over the Internet [22].

2.2 *Serious Games and Edutainment Applications*

Innovations in teaching and learning, promoting a knowledge society, and supporting tremendous growth in demand for highly educated and skilled individuals, are influenced by the advances in technology as well as the emergence of a new generation for whom technology is an integral part of the learning process [3].

Recently, new computerized tools known as 'serious games' (SGs) have appeared in the educational games market. By combining gaming and learning, SGs represent a new area of interest in the educational field [14]. Serious Games and Edutainment Applications arose from the First International Workshop on Serious Games Development and Applications at the University of Derby in 2010 [19]. Over the past decade, there have been tremendous advances in entertainment computing technology and the gang industry grew enormously ranging from a console, PC and mobile based- games [11].

2.3 *Categories of Serious Games*

Serious games could belong to more than one category and the following four are the most common: Military games, government games, educational games, corporate games, and healthcare games.

2.3.1 Military Games

Military simulations have been, and still, are dominant. A board game originally designed for fun can be used in a military training context to teach strategic thinking and the principles of tactical warfare. While the learning process takes place via the game [7]. These games with simple rules, allowing officers to become better planners for battles, have evolved into extremely complex simulators for tanks, helicopters, group training, etc. [22].

2.3.2 Government Games

Training and simulation within the government range from a municipal level to a national level. Governmental games may concern a number of different kinds of tasks and situations, like different types of crisis management, for instance, dealing with terrorist attacks, disease outbreaks, biohazards, health care policy issues, city planning, traffic control, firefighting, budget balancing, ethics training, and defensive driving [22]. Serious games can also be used to games to support cultural heritage purposes, such as historical teaching and learning, or for enhancing museum visits [2]. Win the City game can be categorized under this type of serious games.

2.3.3 Educational Games

For a long time, educators tended to ignore computer games as a source of education. Nowadays, however, the role of games in education is increasing [24]. Games have demonstrated to provoke active learner involvement through exploration, experimentation, competition and co-operation.

They support learning because of increased visualisation and challenged creativity. They also address the changing competencies needed in the information age: self-regulation, information skills, networked co-operation, problem-solving strategies and critical thinking. Importantly, games have become widely adopted by new generations of learners, the so-called digital natives, who have grown up immersed in new communication technologies [25]. Serious games can be used for different types of learning. The nearest at hand are problem-solving activities. Similarly, the player/learner use them to identify with a character so that s/he knows the use of the playing context in real life. Serious games also help in adapting the teaching process according to the learner's profile [15]. Because Win the City game provides a kind of educational exercises for the players to learn about new sites and buildings, it can be also categorized under this type of serious games.

2.3.4 Corporate Games

The corporate games market looms as one of the most lucrative of all other mentioned categories. Games can serve diverse industries and be in forms of small digital games that are designed to advertise some products and services [10]. One field that is constantly evolved through the adoption of new techniques and technologies, especially by incorporation of IT frameworks and techniques is corporate training. Serious Games, uses IT-based techniques, is allowing corporations to improve their training of employees, both by engaging them more actively and by testing their comprehension of the information conveyed [4]. Serious Games introduce different types of applications in corporate settings. Examples of this could be teaching employees to solve problems in a non-traditional way using trial and error or by developing marketable business skills.

2.3.5 Healthcare Games

Serious games applications related to health and healthcare are becoming more common, and today there exist a large number of them [22]. Serious games applied to medical or health-related purposes are growing rapidly in numbers and in types of applications [13]. Serious games present training environments for disaster situations and mass casualty incidents, including combat care. Alongside the training of crisis management, serious games can be used for training everyday clinical activities and skills for junior doctors, e.g. decision-making abilities in surgical procedures or burn-patient care [12].

3 Similar Games

The use of AR and location-based techniques have been investigated for a number of years to develop games that can be used for various purposes. In 2010, two interesting AR games known as ARPuzzle and ARBreakout were developed for educational purposes [18]. Recently, a number of popular location-based augmented reality games have been developed and become very popular. For example, Pokémon Go is a good a AR game example that has a novel social and entertaining experience. Users playing this game not only entertain themselves but gain some health benefits when physically walking trying to locate targets [1, 16]. Another example of a location-based AR game is the social AR serious game that is called Pathomon. This game uses "mechanics of Pokémon GO with the benefits of using AR in a serious context". Pathomon game allows users to scan QR codes and enables different players to team together to achieve one goal [21].

4 Methodology

4.1 Win the City Planning Situation

The City of Valletta has a number of buildings that are not usually discovered and visited. The local council in Valletta aims to offer more information about these buildings and landmarks that are not visible places for visitors. Such information will be provided by developing a serious game that requires the player(s) to be physically in Valletta to be able to play the game which will lead to finding the concerned building.

4.2 Elements of the Game

The main elements of the Win the City game are the environment of the game, the objects included in the game, the goals, the rules of the game, and the player(s).

4.2.1 Environment

The game at this stage focuses on the location of Valletta with the possibility to allocate more cities in the future. Win the City includes only Valeta's map with all utilities and structures including buildings, streets, bridges, and landmarks. The game will not offer a signing in function unless identifying the player's position within a certain range (Valletta).

4.2.2 Objects

The game has a variety of objects, for example; Valletta's buildings, players, building of the day event, military rank level, and the budget. The buildings are three different types; occupied, non-occupied, and landmarks. The players are two teams distinguished by wearing black for the first team and white for the other.

Their level will be counted in gold points and their military level will be counted in the rank for instance; Soldier, Captain, Minister, or King. Each level requires a certain gold scores and procession of buildings. It is important to note that Win the City game identifies one building every day and makes generous offer to encourage players to visit and occupy that building. Such a daily offer puts a higher value of points than the other buildings.

In most cases, these buildings are identified based on the fact that they are the least visited ones. The Puppet Master will be responsible for choosing which building to be chosen every day and players will be updated accordingly on daily biases. Finally, the Puppet master will send instructions during the game in order to offer, explain, and show how to do play and that will be in a shape of balloons appears on the mobile's screen. Figure 1 shows a sample of a Puppet master notification.

Also, the game will represent some paid advertisements about restaurants, hotels, bars, etc. for the player as soon as they pass near them on their way to attack new building. Figure 2 shows an advertisement sample for MacDonald's.

4.2.3 Goals

The main goal of the Win the City game is to discover the city of Valletta e.g. discover the cultural dimension, cultural diversity, and identify the city's history by visiting the buildings and get information about them. Also, the game will allow users to visit the hidden landmarks in Valletta. On the other hand, the game aims to create an

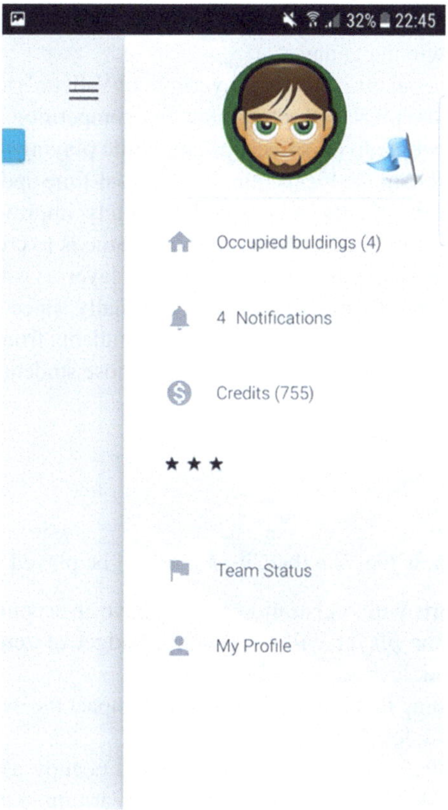

Fig. 1 A sample Puppet Master notification

Fig. 2 Advertisement can be easily integrated as part of the game

entertaining and competitive environment where the players start to challenge each other and attempt to win the game.

Another important goal is to educate players about Valletta's history especially the new generations and the tourists through play and competition. Another interesting goal is to encourage people to do some walking while playing the game to keep fit.

Also, the game will help the local council to get real-time updates about any road, buildings, and landmarks faults, which will definitely improve the infrastructure services in the city. Another goal of Win the City game is to create communication channels between the natives (city residents) and the players (visitors) by having some conversations and answering player's questions. Finally, since Malta is considered as one of the most popular tourist destinations for students from different countries to learn English, the game will help and encourage those students to read, speak, and entertain while playing the game.

4.2.4 Rules

This section defines how the Win the City game will be played.

- First, the player starts with registration steps to have an account to obtain a unique user name. Then, the player will start with a budget of zero gold points and a starting military rank level as a Soldier.
- Second, every building that the player takes will impact the overall budget amount and the military rank level as well.
- Third, the player can play the game and aim to occupy as many buildings as possible in order to help the team collectively or earn more gold pints to win the game individually.
- Fourth, the player can either follow the team's strategy or follow his own plan(s). The player will be ranked individually and as a member of the team. Figure 3 shows the player rankings page and the teams ranking page.
- Fifth, the occupied buildings by one of the teams will not be re-activated on the map for 24 h. However, the other team can attack these buildings and occupy them by bombing them twice and the used bombs will cost double the original price. It is possible that more than one player in the same team can launch the attack.
- Sixth, it is an important rule that players avoid being in the explosion area and be at least 50 m away from any explosion source. In case the player remains in less than that range; the player will be affected and will lose some energy.
- Seventh, the player has limited heavens to use to hide from natives in case the player bombed an occupied building.
- Eighth, the player has to capture as many buildings as possible through 'bombing'. Capturing buildings that had not been occupied before yields better scores. Thus, the inventory 'scores' of the player will be updated based on the number of bombed buildings.

Fig. 3 Ranking of both the players and the teams as part of the gamification engine, which boosts competition amongst different users

- Ninth, an important rule is that the player should know the difference between the types of buildings to avoid being attacked by natives when attacking an occupied building.
- Tenth, another rule is about the landmark. The sculptures and the other landmarks need extra bombs and extra effort to take one of them.
- Eleventh, to be able to attack and occupy a building, the player needs to take a photo for the building by holding the capturing button on the mobile phone for about ten seconds after which the bombing function becomes active. Also, the players have to find a fault in the building to be able to capture it. For example, a broken glass, a hole in a street, damage in the structure, and rubbish at or around the building. Figure 4 shows the capturing function.
- Twelfth, at the end of every month, the Puppet Master will allocate a small number of gold scores to the team that has more buildings.
- Thirteenth, if the player was captured by natives then the player has to go to close heaven to start the game again.

Finally, once a team occupies 90% of the buildings, the Puppet Master will then send a written confirmation of successfully completing the game and winning the city. The winning team will also receive, on their screen, symbols of the city Key and showing the flags of the team on all the buildings on the map. In addition, the player

Fig. 4 The user has to prove that he went next to a particular building and this is achieved through the capturing function of the app

who achieves the highest score will get two free VIP tickets for the New Year's Eve concert and a chance to have a free tour sponsored by the Valletta Local Council.

4.2.5 Player

The game is designed at this stage for two teams players black and white factions and a Puppet master to control the game. The game suits all different ages. There is an interaction between players in the same team; however, there is another interaction between the two groups in a form of identifying the occupied buildings, building that are under attack, and the bombing process. The player can interact with the game environment and use it in a dynamic way. The players can learn about the infra-structure of Valletta and their functions also can learn about buildings, streets, history, and culture of Valletta.

The player can contact a colleague in the same team, identify their case e.g. online or offline, and explore their position on the map. Figure 6 illustrates the flow chart of play and the possible decisions available to be made by the player. At first, the player will be asked to register in order to have a unique user name. Then the player can

Fig. 5 Win the city home page

first watch a short video that explains the main game functions and a sample case of how to play the game. It is important to note that once the player chooses one of the teams, it will be impossible to change to another team unless creating another account and starting all over again. Figure 5 shows how the Win the City game home page look like.

As it can be seen in Fig. 6, the player can build a strategy for play. Also, immediately after successfully signing in into the game, the player can choose from the following possible strategies: finding the location of all the landmarks on the map and decide to attack the available one. Another strategy could be by checking all the non-occupied buildings on the map and capture one of them e.g. the closest, the building of the day, and any non-occupied available building. The last strategy is that the user can try to take one of the occupied buildings and again it is up to the player to choose the suitable choice every time. Figure 7 shows the different strategies the player can use.

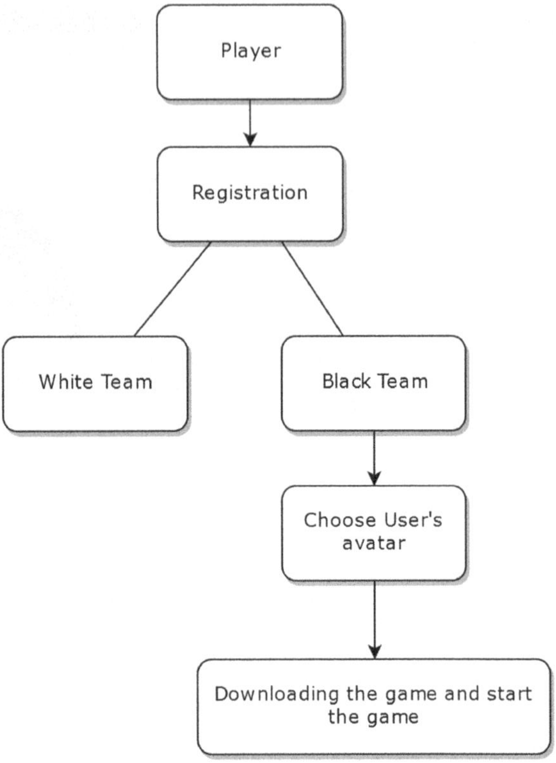

Fig. 6 The diagram highlights the main elements of the game

4.3 Digital Model of the Game

At this stage of the project, a simple interacted prototype was developed. The prototype provides a map of Valetta with a pin representing the player's current location. HTML, CSS, and JavaScript were used to develop the prototype and different codes were used in order Google maps and the GPS service to the game. The model still at the beginning stage but it works well so far. Figure 8 shows the project in Unity.

The prototype at this stage has the ability to connect with Google Maps and choose the type of map's preview e.g. Road Map, Satellite, and also it changes the character position every time the player moves.

In order to manage users' different data, Firebase will be used to create a database which allows to store and synchronize data across multiple clients. Firebase is a full suite for app development that has a cloud messaging called Firebase Cloud Messaging (FCM). FCM is a cross-platform solution for messages and notifications for Android, iOS, and web applications, which currently can be used at no cost. A new account for the project was created and an Android SDK was downloaded to be added to Unity project. This SDK will allow users to verify their registration through their email addresses and passwords and using this information to sign in to enter the game. Another feature of using Firebase that user can reset their passwords by emails.

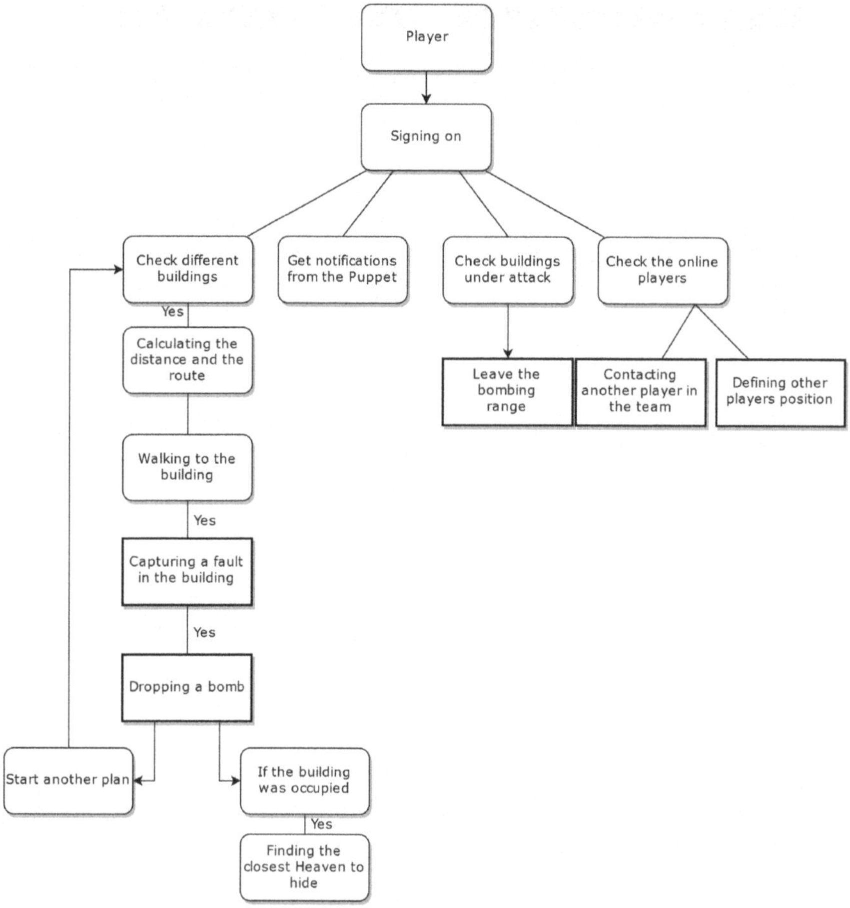

Fig. 7 The user can adopt different game playing strategies throughout the game

Also, with this SDK users are allowed to sign in with their Google, Facebook, Twitter, and GitHub accounts. Finally, during the game Firebase will allow the application to use a cloud and connecting in the real time to the database. Figure 9 Win the City project on firebase.

4.4 Future Considerations on the Digital Model of the game

A number of saved Google maps were used to simulate the different scenarios of the game. For instance; the current player's position, the position of the other players from the same team on the map, the position of the other team's players, different markers to identify different buildings, special markers to identify the status of the

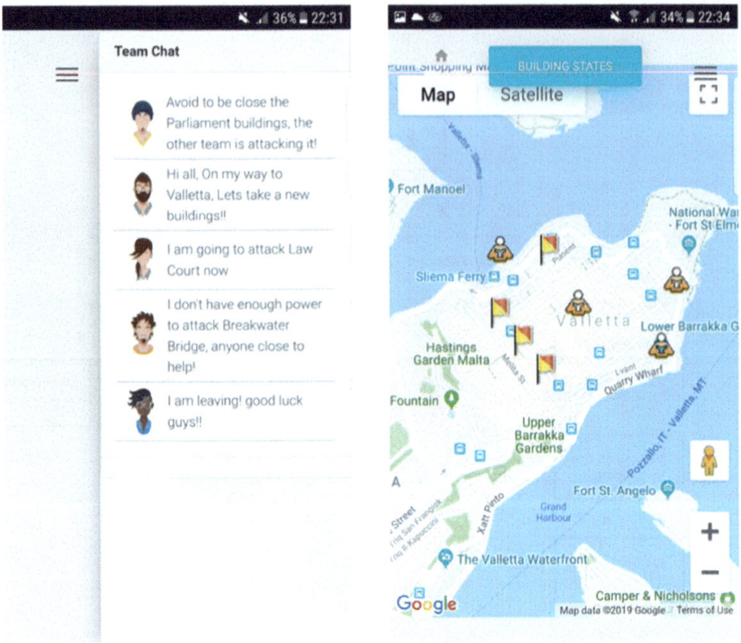

Fig. 8 Different aspects of the game showing the group chat and the location map

buildings on the map, and possible routes to reach a chosen building. In addition, some draft maps were created to simulate the bombing process and the explosion range calculated in meters and represented by a designed symbol.

Also, the prototype allows users to interact by touching an action mark of a player, a building, and a landmark symbol on the map in order to get a notification message about any of the game's elements.

5 Experiments and Testing

This stage was completed through the following three main steps; after a first version of the prototype was developed, the prototype was installed on a Samsung s6 edge mobile, and a number of users tried the game. Information was gathered from the users before trying the game prototype and after using it.

The gathered information before testing the game included: users' demographic information, users' previous knowledge and experience with serious games and playing them, willingness to play serious games about sites/buildings in Valetta including rewards preferences, and type of mobiles devices the users have and whether they have Internet connections on these devices. After the users tried the prototype, 13 questions were asked. The first 12 questions used Likert five-point level of agreement

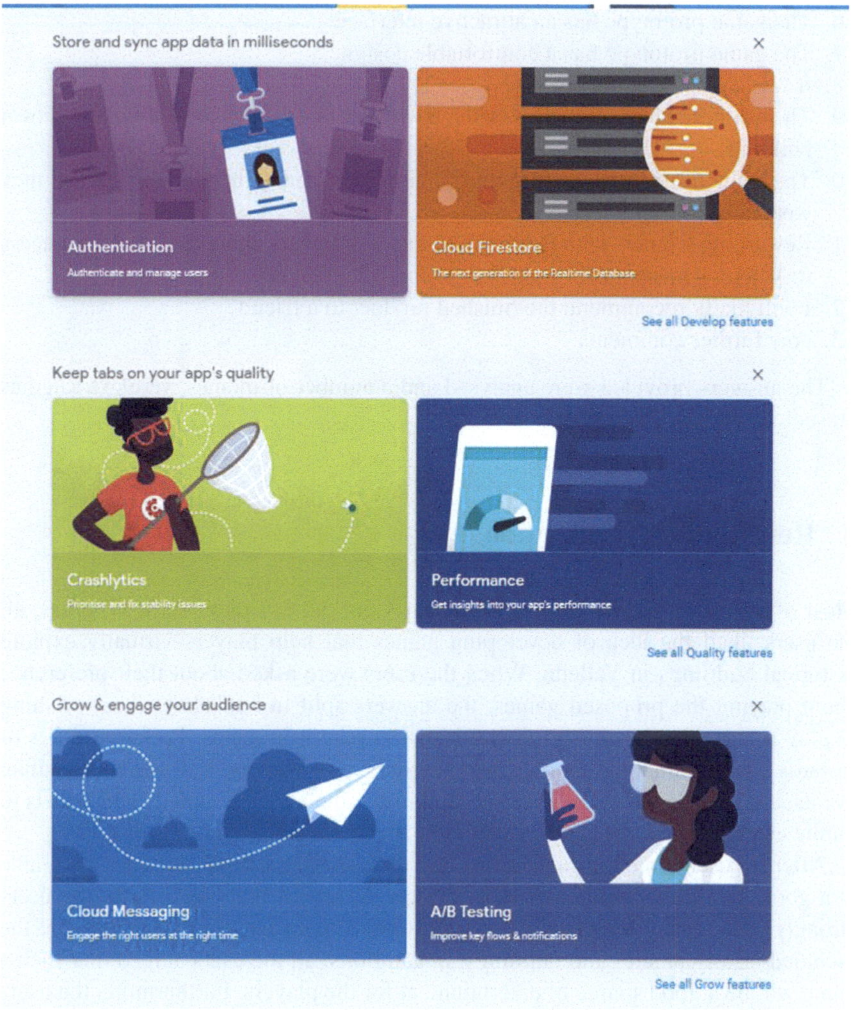

Fig. 9 The firebase project

scale (Strongly disagree; Disagree; Neutral; Agree; Strongly agree). These questions were as follows:

1. Win the City game is a good way to encourage users to explore historical buildings
2. Win the City game will assist tourists to easily navigate Valletta's historical sites
3. Win the City game will be a good source of entertainment for the users
4. The game will provide valuable feedback to the local council
5. The game will indirectly benefit the players' health when they walk to reach different buildings

6. The game prototype has an attractive interface
7. The game prototype has a comfortable design
8. It was easy to use the prototype
9. The example I have seen provided me with enough information about the shown building
10. The needed functions to play this game where placed in a clear place and they were easy to use
11. Rewarding winners with free public event VIP tickets, monetary, and free access to Valetta's museums is good way to motivate players
12. I will likely recommend the finished product to a friend?
13. Any further comments

The answers provided were analysed and a number of themes were extracted as presented in the following section.

6 Results of the Evaluation

Most of the users heard about serious games but did not play them. Whereas, all the users liked the idea of developing games that help players virtually explore historical buildings in Valletta. When the users were asked about their preference about playing the proposed games, the answers split in half about those wishing to play it individually and those who stated to play it in teams. Preferred types of rewards when winning the games were mixed e.g. monetary, VIP tickets to public events, and free access to Valletta's museums but the majority stated 'VIP tickets to public events'.

After the users tired the game prototype, the users agreed that Win the City Game is a good way to encourage users to explore historical buildings. Also, the users strongly agreed that this game will assist tourists to easily locate and navigate the included historical sites and buildings. In additions, all the users agreed that such a game will be a good source of entertainment for the players. Furthermore, the users agreed that such a game will provide the local city council with valuable feedback about the buildings and sites included in the game. When the users were asked about if playing the game would enhance the players' health as a result of walking from one physical site to another, they had mixed responses some of which were; neutral, agree, and strongly agree.

When the users were asked about the attractiveness, design, and ease-of-use of the tested prototype, they all strongly agreed that it was attractive, has a comfortable design, and it was easy to use. In the same line, the users agreed that after playing the game, they have better information about the site/building used in the testing round.

From the above, we believe that the developed game is a novel way to make potential users aware of the existent of some historical sites and buildings in the first place. We also think that rewarding the winners with some of the mentioned awards will ask as a direct motivator for potential players to play the game. Furthermore, we expect the popularity of the game to grow fast because all of the users who tested the prototype stated that they will recommend the game to people they know.

7 Conclusions and Further Work

Win the City game aims to bring knowledge and pleasure to the process of interaction between different players and between players and residents of Valletta city. In this project, the focus was on designing a serious game that will help players to discover famous historical sites and buildings in the city of Valletta and at the same time acts as a source of entertaining. The main goal was to design and develop a mobile, public participatory, and competitive serious game and we believe we managed to fulfil this goal. However, there might be some high investment costs and complexity of creating a mobile game using Augmented Reality technology and we recommend that Android Studio is used to develop a similar game. It is recommended for future research to allocate more resources and time for a similar project.

Also, it is wise to consider involving experts from education, entertainment, and game design areas to develop a similar game in the future so that more functions are included, and perhaps better rewards are offered. Finally, researchers may consider including more than two teams in the game in order to toughen the competition, allow more people to play the game, and make winning the city take longer.

References

1. T. Althoff, R.W. White, E. Horvitz, Influence of Pokémon Go on physical activity: study and implications. J. Med. Internet Res. **18**, e315 (2016)
2. E.F. Anderson, L. McLoughlin, F. Liarokapis, C. Peters, C. Petridis, S. de Freitas, Serious games in cultural heritage (2009)
3. S. Arnab, T. Lim, M.B. Carvalho, F. Bellotti, S. De Freitas, S. Louchart, N. Suttie, R. Berta, A. De Gloria, Mapping learning and game mechanics for serious games analysis. Br. J. Educ. Technol. **46**, 391–411 (2015)
4. A. Azadegan, J.C. Riedel, J.B. Hauge, Serious games adoption in corporate training, in *International Conference on Serious Games Development and Applications* (Springer, Berlin, 2012), pp. 74–85
5. F. Bellotti, R. Berta, A.D. Gloria, L. Primavera, Enhancing the educational value of video games. Comput. Entertain. (CIE) **7**, 23 (2009)
6. E. Boyle, T.M. Connolly, T. Hainey, The role of psychology in understanding the impact of computer games. Entertain. Comput. **2**, 69–74 (2011)
7. J. Breuer, G. Bente, Why so serious? On the relation of serious games and learning. J. Comput. Game Cult. **4**, 7–24 (2010)
8. T.M. Connolly, E.A. Boyle, E. MacArthur, T. Hainey, J.M. Boyle, A systematic literature review of empirical evidence on computer games and serious games. Comput. Educ. **59**, 661–686 (2012)
9. D. Djaouti1&2, J. Alvarez, J.P. Jessel, O. Rampnoux, Origins of Serious Games (2012)

10. D. Djaouti, J. Alvarez, J.P. Jessel, Classifying serious games: the G/P/S model, in *Handbook of Research on Improving Learning and Motivation Through Educational Games: Multidisciplinary Approaches* (IGI Global, 2011), pp. 118–136
11. S. De Freitas, F. Liarokapis, Serious games: a new paradigm for education? in *Serious Games and Edutainment Applications* (Springer, London, 2011), pp. 9–23
12. M. Graafland, J.M. Schraagen, M.P. Schijven, Systematic review of serious games for medical education and surgical skills training. Br. J. Surg. **99**, 1322–1330 (2012)
13. M. Graafland, M. Dankbaar, A. Mert, J. Lagro, L. De Wit-Zuurendonk, S. Schuit, A. Schaafstal, M. Schijven, How to systematically assess serious games applied to health care. JMIR Serious Games **2**, e11 (2014)
14. C. Girard, J. Ecalle, A. Magnan, Serious games as new educational tools: how effective are they? A meta-analysis of recent studies. J. Comput. Assist. Learn. **29**, 207–219 (2013)
15. N. Hamdaoui, M.K. Idrissi, S. Bennani, Serious Games in Education Towards the standardization of the teaching-learning process. *Advances in Educational Technologies*, 174 (2014)
16. J. Koivisto, A. Malik, B. Gurkan, J. Hamari, Getting healthy by catching them all: a study on the relationship between player orientations and perceived health benefits in an augmented reality game, in *Proceedings of the 52nd Hawaii International Conference on System Sciences* (2019)
17. F. Laamarti, M. Eid, A.E. Saddik, An overview of serious games. Int. J. Comput. Games Technol. **2014**, 11 (2014)
18. F. Liarokapis, S. De Freitas, A case study of augmented reality serious games, in *Looking Toward the Future of Technology-Enhanced Education: Ubiquitous Learning and the Digital Native* (IGI global, 2010), pp. 178–191
19. M. Ma, A. Oikonomou (eds.), *Serious Games and Edutainment Applications*, vol. 504 (Springer, London, 2011)
20. L. Michaud, J. Alvarez, Serious games. *Advergaming, Edugaming, Training... IDATE Consulting & Research* (2008)
21. D. Rapp, J. Müller, K. Bucher, S. von Mammen, Pathomon: a social augmented reality serious game, in *2018 10th International Conference on Virtual Worlds and Games for Serious Applications (VS-Games)* (IEEE, 2018), pp. 1–4
22. T. Susi, M. Johannesson, P. Backlund, Serious Games: An Overview (2007)
23. B. Sawyer, Serious games: improving public policy through game. *Foresight and Governance Project* (2001)
24. L. Stege, G. Van Lankveld, P. Spronck, Serious games in education. Int. J. Comput. Sci. Sport **10**, 1–9 (2011)
25. W. Westera, R.J. Nadolski, H.G. Hummel, I.G. Wopereis, Serious games for higher education: a framework for reducing design complexity. J. Comput. Assist. Learn. **24**, 420–432 (2008)

Reversing Heritage Destruction Through Digital Technology: The Rekrei Project

Marta Massi and Alfredo D'Angelo

Abstract Digital technology has paved the way to a progressive disintermediation in many industries including the arts and cultural sector. This chapter examines the role of digital technology as a means to reclaim the heritage that is being lost or destroyed, by illustrating the entrepreneurial case of Rekrei. Building on the disintermediation and value cocreation literatures, this chapter highlights how digital technology can change the way heritage is funded, produced and consumed, by enhancing customer participation and value co-creation. Managerial and policy implications emerged from this research.

1 Introduction

This chapter examines the role of digital technology as a means to reclaim the heritage that is being lost or destroyed by illustrating the Rekrei case, an entrepreneurial initiative that allows for virtually reversing heritage destruction. Developed by Matthew Vincent and Chance Cougheneur in 2015, Rekrei was originally named Project Mosul and aimed at virtually rebuilding the Museum of Mosul that was destroyed as a consequence of the attacks by the Islamic State of Iraq and Syria (ISIS). Owing to the success of the initiative, the project was renamed Rekrei, an Esperanto word meaning 'recreate', and extended to any cultural heritage around the world that has being destroyed or dispersed due to wars or natural disasters. Crowd-funding was used to raise capital for this entrepreneurial idea.

Building on disintermediation, value co-creation [20, 21] and Service Dominant Logic [25] the Rekrei case shows how digitalization, acting as a means to rediscover

We thank Costanza Piazzolla for contributing to the data collection.

M. Massi (✉) · A. D'Angelo
Universita Cattolica del Sacro Cuore, Milano, Italy
e-mail: marta.massi@unicatt.it

A. D'Angelo
e-mail: alfredo.dangelo@unicatt.it

D. Seychell and A. Dingli (eds.), *Rediscovering Heritage Through Technology*, Studies in Computational Intelligence 859,
https://doi.org/10.1007/978-3-030-36107-5_6

heritage that is being lost or destroyed, enabled a groundbreaking and innovative collaborative form of production, conservation and consumption of heritage.

The Rekrei case is aimed at pointing to new, innovative, and collaborative forms of heritage production and consumption. The chapter will also highlight how digital technology can change the way heritage is funded, produced and consumed, by enhancing customer participation and value co-creation.

The chapter is structured as follows. First, we present and discuss the theoretical foundations, including disintermediation, decoupling, Service-Dominant Logic (SD-L) and value co-creation. Second, we present the case supported by an interview conducted with Chance Coughenour, one of the founders of the project. Third, we report and discuss the main findings and implications, presenting our concluding remarks, limitations and avenues for future research.

2 Literature Review

This section reviews the literature streams that the authors employed to frame the Rekrei case. In particular, the theories of disintermediation [14], decoupling [18], Service-Dominant logic [25] and value co-creation [20] will be employed in order to analyze and discuss the Rekrei project.

2.1 Disintermediation

The advent of digitalization has had disruptive effects on many organizations, leading to a progressive disintermediation of their activities and processes. In fact, digitalization has brought about critical changes pushing businesses in different sectors to re-envision their traditional business practices and models [19].

The phenomenon of 'cutting out the middleman' [14], initially introduced in the finance sector to indicate the elimination of financial intermediaries (such as banks and brokers), has gradually expanded its effects in many other sectors ranging from tourism to intellectual properties and creating entrepreneurial ecosystems that overstep the notions of geographical space and physical presence. Thanks to the advent of the Internet and new digital technologies, companies can now deal with every customer everywhere in the world directly without using traditional intermediation channels such as distributors, wholesalers, brokers, or agents. At the same time, customers can interact directly to exchange ideas, products, services and funds based on a collaborative perspective, which now expands beyond the national boundaries.

The phenomenon of disintermediation is particularly evident in service-based industries, so that many have talked about the advent of the era of the sharing economy. Companies such as Uber and Airbnb, which allow users to share houses and cars, represent typical instances of organizations that have embraced this disintermediation process, which is redefining how products and services are consumed. A process of 'Uberization' is taking place and increasingly many services are offered that empower consumers to take an active role in the delivery process.

The effects of disintermediation have been particularly disruptive also in arts industries including music, print and film, "decimating entire sectors of the media arts industry and altering the landscape for its production, dissemination, and enjoyment" [19]. This shift is even more evident in the visual arts sector where the digital revolution has gradually corroded the aura of originals or prototypes [2].

The heritage sector is one of the sectors where technology has disintermediated the value chain and delivery processes, providing organizations with new entrepreneurial opportunities based on sharing knowledge and contents with their users, independently from their geographical proximity or physical presence. For instance, there is a growing interest of arts organizations, such as museums and theatres, in digitalizing or virtualizing their heritage collections and archives as well as in diffusing educational content throughout the Internet [24]. The notion of virtual cultural heritage has been recently introduced to indicate new ways of consumption of cultural heritage that take advantage of digitalization processes.

As a result, new disintermediation and participatory logics have emerged that allow consumers to "participate actively in the art they consume" [19]. In fact, art audiences do no longer conceive of themselves as mere consumers, but "want to be involved, to engage in the conversation around art and creativity and perhaps participate in its production [19]".

Digitalization has introduced new opportunities for an increasingly more active role of consumers. In particular, 3D acquisition and modeling technologies have been defined as ideal for the recording of heritage sites [5, 10]. This technology allows for generating a "digital simulacrum of a real artifact, and the availability of digital tools for manipulating, exploring, comparing and explaining a virtual object", thus greatly increasing "the comprehension and the valorization of monuments" [11]. Projects such as Cyark, 3D Icons, and ScottishTen, which digitally record, archive and share cultural heritage, are examples of the application of digitalization.

Furthermore, projects such as Curious Travellers, the Million Image Database, Heritage Together and Share Our Cultural Heritage take a step further in that they are not only aimed at the documentation of archaeological sites and monuments but also employ crowdsourced data. Rekrei takes a step further in that it is crowdsourced and crowdfunded, thus empowering consumers who become prosumers in the process of content creation. Rekrei represents, therefore, an innovative paradigm of heritage consumption and production that takes advantage of disintermediation.

2.2 *Decoupling*

A consequence of disintermediation processes is the phenomenon of *decoupling*, defined as the separation of the elements of structure from activities [18]. Increasingly value chains, which were designed and developed under the assumption of the old economy, have been redesigned in response to the effects of digital production, distribution, and consumption [4]. In particular, as a consequence of the advent of the Internet, businesses are gradually decoupling their activities by building extensive networks and providing platforms that allow customers to share their experiences and dialogue with the company and amongst themselves [20].

Firms progressively decouple their structures from their activities by creating "an experience environment in which customers can have an active dialogue and co-construct personalized experiences" [20]. A consequence of decoupling is the creation of dynamic platforms for consumer practice, which allow the creativity and know-how of consumers, and at the same time "channel these consumer activities in ways desired by the firm" [27]. Such engagement platforms and interactive environments are critical for the emergence of customer experiences, which allow individuals to participate actively in the value creation experience.

The emergence of decoupling processes is particularly evident in some sectors, e.g., the music industry [4], which has started to decouple its structures from its activities in order to reduce manipulation by peer-to-peer virtual networks and hackers. With the advent of the digital revolution, music consumers are provided with new value co-creation opportunities such as custom playlists creation, loop sampling, remixing, and audio editing software, which "provide tools which blur the distinction between the artist who creates the music and the user, who once passively consumed it" [15].

Decoupling is increasingly employed by organizations because it allows them to maintain legitimacy "by adopting designs that mask or distract attention from controversial core activities that may be unacceptable to some key constituencies" [7].

Decoupling also relates to the separation of contents, such as text, sound, images and videos, from the "physical materiality of the services and devices used to produce and consume them", introducing the concept of digital materiality [16]. This is particularly evident in the notion of 'virtual cultural heritage' which points to new ways of consumption of cultural heritage that take advantage of digitalization processes. In this chapter, we show how Rekrei takes advantage of decoupling both by providing a platform that allows customers to become producers of contents and by encouraging the creation of a heritage characterized by digital materiality.

2.3 Value Co-creation and SD-L

Value co-creation has emerged as a novel conceptualisation in marketing theory [9] and it is often discussed in relation to the Service-Dominant Logic (S-DL) [25]. Based on the SD-L, companies do not directly create value but can offer 'value propositions' and subsequently create value in a collaborative way with potential users. Under this perspective, value creation is never understood as a solipsistic process. Value can be created only in collaborative and interactive contexts where multiple exchanges occur under a many-to-many perspective. This exchange of value could also take place in a tacit and indirect manner, especially if it is activated starting from a tangible asset [26].

In particular, value co-creation is linked to the idea that the concept of 'value-in-use' has superseded that of value in exchange [17]. Traditional marketing is focused on the firm-centered notion of value in exchange, i.e., making a value proposition that passive consumers can accept or decline [20]. In contrast, the new perspective emphasises the co creation of value by the supplier and the empowered customer. The distinction between "value-in-use" and "value-in-exchange" makes it possible to look at value and value creation in an alternative way, and is critical to understand the shift from a goods-dominant to a S-DL [25]. Value-in-use, which exists for an individual to the extent that a product "directly satisfies his [or her] wants", leads to a blurring of the roles of producer and consumer where value components (the benefits and sacrifices that generate value-in-use) are negotiated through collaborative problem-solving [1].

Introduced by [20], the concept of co-creation represents a novel theorization of the relationship business-consumer. Based on the co-creation paradigm, the traditional firm-centric view is compared to classical theatre where a clear separation between actors (firms) and the audience (customers) occurs, while the customer-centric view is assimilated to experimental theatre in which the roles of the firm and the consumer converge [20]. The consumer actively participates in the creation of products and brands, thus feeling more empowered.

Shifting the focus of marketing thought from a firm-centric to a collaboration perspective [9, 20], value co-creation implies that not only producers but also customers contribute to creating value through individual co-creation experiences and interactions with brands, companies, and other consumers.

Through value co-creation, consumers become increasingly interconnected and are no longer passive actors in the contemporary business scene, always ready to provide feedback to businesses, participating in the ideation of products and services or coproducing them. Co-creation processes have led to the introduction of peer production, which has been defined as "the most theoretically radical organizational innovation that has emerged from Internet-mediated social practice" [3]. Taking advantage of new digital technologies, users are continuously generating and exchanging content with their peers, so that eWOM (Electronic Word of Mouth) and SoLoMo (social-local-mobile) have become new trends, changing the way many products, including heritage, are consumed.

A branch of co-creation practice is crowdsourcing, i.e., the act of a company or institution taking a function once performed by employees and outsourcing it to an undefined (and generally large) network of people in the form of an open call [13]. Rekrei represents an instance of co-creation practice and crowdsourcing in that it allows for building, uploading and downloading 3D models of destroyed heritage. In addition, Rekrei takes a step ahead in that it also allows users to fund the project by donating, thus empowering and engaging users even further.

3 Research Methodology

In order to investigate the case at study, a qualitative approach was adopted. Owing to the exploratory nature of this research, the qualitative method was preferred to provide a rich and in-depth description to analyse the complex and multifaceted nature of the phenomenon in question.

To gain understanding of this case study, data were collected through an in-depth interview with a key informant, one of the Rekrei Founders, Chance Coughenour. Collected data were analysed utilizing an open coding approach [6], to develop meaning categories and themes [12]. The interview was transcribed and examined recursively and separately by two researchers, who identified recurrent themes. Using these categories, the researchers coded the data independently, and input the coding into NVivo 10 as a means of organising the disparate data. After coding the data, the researchers compared their coding and assessed inter-coder reliability.

3.1 Use Case

Rekrei is "an open-source online platform, which provides a simple user interface for identifying locations of destroyed heritage, uploading and sorting images, and a 3D gallery of completed reconstructions" (Chance Coughenour, Founder of Rekrei).

Originally named Project Mosul, Rekrei is a project developed by Matthew Vincent and Chance Cougheneur in 2015. The intent of the project is to give new life to that part of cultural heritage that has been destroyed or dispersed due to wars, natural calamities, or misfortunes of all sorts. Through a database, made available on the Rekrei website, anyone can collaborate to replenish the collection of photographs of damaged monuments, museums or artifacts. The aim is to virtually reconstruct the findings that were physically lost in 3D, in an attempt to preserve the global heritage that we share. Not by chance, the name Rekrei is the equivalent of the verb 'recreate' in Esperanto, an artificial language developed ad hoc at the end of the 1800s in order to break down linguistic limits and create a universal language.

Rekrei employs 3D technology to create digital reconstructions of heritage or artifacts by using crowdsourced images that were taken before the heritage destruction. It takes advantage of photogrammetry, an innovative technology that allows to create a 3D model by using two-dimensional images of the same object taken from different angles [22]. Anyone can participate in contributing photographs of damaged monuments sites or other heritage, museums and artifacts, thus enriching the database made available on the Rekrei website. In fact, participation is crucial because the more the crowdsourced images of an object, the higher the definition of the 3D model that virtually recreates that object.

In addition to the features of other platforms, Rekrei also includes a tool that allows for extracting images from websites such as Flickr, based on their geotags, to complete reconstructions (Chance Coughenour, Co-founder of Rekrei) and a section to donate in order to fund the project. Currently, the database contains more than 6,000 photographs.

3.2 Case Analysis

The Rekrei case has been selected as an example of platform based on decoupling and disintermediation, which encourages co-creative practice, thus representing an instance that exemplifies the aforementioned theories.

3.2.1 Disintermediating Heritage Production and Consumption Through Collaborative Technology

While a paradigm based on intermediation guarantees a linear flow of communication (from the producer to the consumer), disintermediation-based models introduce different communication patterns and directions. Since the communication flow is not conditioned by the presence of the intermediary, a multidirectional communication is more likely to occur. This shift in communication also affects value creation processes, which are no longer linear. Instead, a value constellation paradigm emerges that allows anybody to contribute to value creation. Peer production means that there are no longer defined roles of producers and consumers, even when it comes to heritage [3].

In this sense, Rekrei is a great example to illustrate the effects of disintermediation on the production and consumption of heritage. From its embryonal phase (when the idea was conceived on Facebook) to the current state of the project, Rekrei has been the consequence of a non-linear communication flow and a value constellation to which different actors have contributed.

In February 2015, when the video of the destruction of the Mosul Museum became viral, a young community of archaeologists was ready to react. Starting from the Facebook group of the ITN-DCH project—which promotes and creates technological innovation in the field of digitization of cultural heritage—a video post was published to call for action. From the debated emerged from that post, as Chance Coughenour, one of the founders of Rekrei, stated, the idea of the former Project Mosul was born:

Matthew and I were on the ITN-DCH project at the time and we had a Facebook-page for the ITN-DCH project for us to communicate what our project was doing with different Ph.D.-students from across Europe. The video of the Mosul Museum destruction was of course viral, I think it was on February 26 or 27 in 2015 and Matthew posted the video to our Facebook-page to communicate with people who were following our project because it was related of course to digital cultural heritage in general and, you can find this post if you scroll back through our ITN-DCH Facebook page. It's a long acronym ITN-DCH (Initial Training Network—Digital Cultural Heritage). So, if you scroll back to the post, back to the 2015 February you will find that Matthew posted that video and when I saw that posted, I replied in the comment section and I said, you know: "It's a terrible thing", and Matthew was communicating this about how this is a terrible thing to see heritage been destroyed and I think my comment was something about: "Maybe we could find some photographs that people had taken from before these artifacts were destroyed, and maybe we can create some form of a partial reconstruction 3dimentionally and hopefully even if we hadn't enough of these reconstructions we can put them back inside a virtual museum to re-tell that story", that was the comment that came from Matthew's post, and then he responded and said: "I think it would be amazing to do that: to reconstruct what ISIS was deconstructing" and so, it's a good example of how social media was the actual catalyst for ... because, then we had other Ph.D.-students reading the post, other people who were following our page reading the post and liking these comments and then we started to think about it, even later that day ... "I think we can actually do this!"

Before being an independent platform, Rekrei was developed through social networks. Social networks allow individuals to be producers of contents, stories, ideas, and services. Social networks can activate bonds and relationships, completely breaking down barriers or distance, and disintermediating content production processes. Coughenour is clear about the importance of social networks in the development of the Rekrei project:

The initial inception of the concept of project Mosul and Rekrei came from social media and built on social media, so, it's actually a product of the social media. (...) I can see the potential is immensely, it's eminent, huge (...) I would say that it's extremely important as it can be used in different ways to share 3D models or to share stories that are related to the project.

Social networks introduce collaborative technology that also allows for solving problems. Since its creation, the project increasingly expanded the connections of its network to find users who could deal with that particular type of virtual reconstruction. Chance highlighted how he found a solution thanks to the relationships established within the ITN-DCH research group:

The first challenge was: we needed a web-site, luckily Matthew was able to build one very quickly based on our conversations on what we would need but; apart from the website creation we also had the challenge of figuring out, because from our understanding this had not really been done before, the only other examples were what Fabio Remondino and some people had done in Zurich with the Bamiyan Buddhas. When the Bamiyan Buddhas were destroyed in 2001 Fabio Remondino, who was and is one of our advisors for our project, was also working on the ITN-DCH project, so you see there are a lot of connections here that make a lot of sense.

The actual version of Rekrei makes of disintermediation its main characteristic, so that anyone can contribute to the process, by uploading or downloading images of lost heritage, thus acting at the same time as producers and consumers:

They may have photographs of heritage that was destroyed, they may know someone who had visited one of these sites so their own contribution from that point of view is they learned about the project and then they help disseminate and grow the network. (...) they know how to use photogrammetry so they can download the photographs and create 3D models and upload them to the website via Sketchfab.

In practice, Rekrei represents an innovative heritage conservation project that takes advantage of disintermediation and peer production:

We were trying to use crowdsourced, volunteer and tourist photographs which are different and we were also building a platform so people can download and upload 3D models whereas this form of dissemination and output was through a research article, back then.

3.3 Decoupling for Legitimacy

The decoupling process can be read as a business attempt to facilitate consumers' co-creative practices while at the same time maintaining control over the process. Thus, value co-creation obtained though decoupling can be assimilated to the concept of governmentality, which refers to power as a form of social control, rather than a hierarchical, top-down form of control, "which is aimed at shaping the actions of individuals through the imposition of orders, rules, and norms" [8]. In contrast, governmentality "works from the bottom up and represents a form of power that acts through practices that 'make up subjects' as free persons" [23]:

I think Rekrei represents a project or concept that allows people to feel empowered and then build public awareness by utilizing and using, basically crowdfunding for the creation or re-creation of destroyed heritage which helps grow the understanding of culture across the global community.

By constructing consumers as partners in mutually beneficial production processes, firms retain control over the value co-creation processes. Decoupling is usually employed as a strategy to avoid product and brand devaluation by making products and brands appear as authentically co-created or distant from commercial considerations, thus maintaining legitimacy and support from the constituencies. The same logics apply to the Rekrei case:

It needs to be unconstrained by economy, it needs to be unconstrained based on why they are doing it, and that means like you said before it's about the passion that you have and the interest that you have to do it. So, that's important because if you are just trying to do it for money, it's going to become difficult, because you are not going to get as many people interested, because if you are the only one who is going to receive economic gain, why would volunteers be working for you to make money.

The separation of the Rekrei Project from economic considerations, promoted by the presence of the crowdfunding initiative, also contributes to enhance the perception of the project as authentic. 'Authentic' is synonymous with disinterested: "To be authentic, brands must be disinterested; they must be perceived as invented and disseminated by parties without an instrumental economic agenda, by people who are intrinsically motivated by their inherent value [12]".

3.4 Turning Heritage Consumers into Heritage Producers: A Co-creation-Based Perspective

Removing the 'middleman' makes the communication flow more fluid and leads to a blurring of the roles. Postmodern scholars refer to a reversal of production and consumption, and prosumption in which the roles of suppliers and users are increasingly blurred.

The co-creation of value is facilitated by the Rekrei platform that allows the consumer a constant and widespread updating. This update is not an end in itself since data, ideas, information accumulate, increasing knowledge. Thanks to the exchange of knowledge that consumers will be able to collaborate in the creation of value. Rekrei is a good example of how consumers can collaborate in the activation of development and innovation processes in the context of arts and culture. As Coughenour pointed out:

It is not one person creating anything, it's always a question of a lot of people coming together to create and build a product, or platform or some type of project or object.

The value is co-created by all the actors who participate in the project, including sponsors, partners and public institutions based on a value constellation paradigm:

More and more photogrammetry companies will contact us and learn about our project through conference papers or press stories, and they would like to be a part of it too, so we basically partner with corporate partners only if they will provide us with a license for our volunteers, so volunteers don't have to pay as long as they know to use the software ... that's more than just the license providing. I think the value is instead that we can list them as a partner on our website which helps to legitimize and saying you know: 'Look, these companies find value in supporting us in some way', not particularly financially but they are at least providing software for our volunteers and I think the companies ... you know, what value they got out of it is number one: they love the idea of the project and number two: it gave them additional press, if they chose to pursue it, they would at least be mentioned ... or when people came to our website they would see their company listed too, so there are mutual value both ways.

One of the main partners of the Rekrei project is the Economist that in 2015 financed the creation of the virtual museum of Mosul, RecoVR: Mosul, and the 3D printing of the lion of the Mosul museum, which has been on display in several museums, including the Venice Biennale:

The Economist, I guess is a corporate partner too. In their case, they wanted to experiment with VR storytelling in a way they had never done before. They never did this before and really only maybe the New York Times was beginning at that time to explore like 360 storytelling with VR. So, I think The Economist found it as the perfect opportunity to say: 'Hey, we don't want to like, replace what you are doing but we want to help you achieve the story that you can't do because you're not being funded' so they helped us fund the creation of the Virtual Museum, they funded the printing of the 3D objects like one that you may have seen in my Ted-x talk and these

3D objects when they were printed we were displaying them in Museums or at the Venice Biennale and this was all funded from The Economist.

The network created by Rekrei and the derived resonance is an extremely positive consequence of the value co-creation. The value chain is solid and based on the resources that each single node can transmit to the next node. Everyone in the network shares a common goal: the preservation of the lost historical and artistic heritage. In this context, the value chain is made up of various actors and diverse communities, i.e., academic, technological, media and press stakeholders, who spread the word about the project to eventually get to the general public. These actors, each with his/her own contribution, make it possible to achieve the main goal of Rekrei:

I think the most important network is the really simple one, it's actually one of the many at the basics sense ... they are the connection between: heritage community, history community, or archeology community with technology and not just applying technology to archeology project or history projects but actually connecting those two things ... so you have ancient art, or ancient heritage and then you have technology that can be used to produce this types of story and then connecting it to the general public. So, and the way that this is connected to the general public is through the media and the press. This network is a kind of move back and forward and without the media, without the press we wouldn't have been able to have this many general public, people learn about our project.

Based on the SD-L, the result is a service that is balanced by the integration of the resources of every actor and enhanced by each one's contribution. The collaborative and interconnected nature of value creation makes is possible that all partners, who intend to join the network, can benefit from it in order to achieve a balanced centrality. This means that to build long-term trust-based relationships and well-functioning markets, the needs and wishes of multiple stakeholders—including customers, employees, suppliers, intermediaries, governments, the media—should be taken into account.

In sum, the upside of the project is to make individuals become co-creators and active participant in a story that is universally important:

What makes Rekrei kind of unique is that it fits between things that are done in the research community, in Academia, connecting them to a public story, a story that the public is following online or on TV about what is happening in the world to global heritage and it kind of connects that gap. (...) Rekrei was like an action-project it wasn't something that someone would read a press story and say: 'Oh, that's interesting what you're doing'. No, they can read the press story and say: 'I can join that.' (...) they recognized that they could help do something-even though they felt they couldn't about heritage been destroyed.

4 Discussion

Rekrei represents a groundbreaking heritage conservation paradigm in that it provides destroyed heritage with new life, by creating a 'virtual museum' that preserves the memory of lost heritage and makes it possible to hand it down to the next generations. Not only does Rekrei represent an innovative take on virtual reality, but it also presents an interesting paradigm for crowdsourcing and collaboration of different actors, including individuals, public institutions and private industry.

The Rekrei case emphasizes how interactive platforms and environments could be critical for the emergence of customer experiences, which allow individuals to participate actively in the value co-creation experience. Cutting out the middlemen has a number of consequences, of which the main seems to be the loss of control by institutions which are traditionally designated to manage heritage and artworks, i.e., government, museums or galleries. Such institutions assert control over specific processes and contents. Their legitimacy, i.e., being able to conform to accepted norms, provides them with the right to operate and manage the matters that they are responsible for. Under a decoupling perspective, the Rekrei case shows how museums and art galleries may not be the only institutions who are able to display artworks or heritage objects and opens up new opportunities for public and private organizations that would be willing to work in this sector.

Managers should look at Rekrei as an example of project that allows for cultural, social and economic sustainability. By providing cultural heritage that has been destroyed or dispersed due to wars, natural calamities, misfortunes of all sorts, with new life through digital restoration, Rekrei guarantees cultural sustainability. In this way, it works as a cultural guarantor, breaking down the local and cultural boundaries and functioning as the 3.0 custodian of the global cultural heritage.

In doing so, Rekrei fulfills the second axiom of the S-D Logic: "The consumer is always a co-creator of value" in that it allows its users to join different types of interaction. Co-creation occurs by uploading photos of lost or destroyed artifacts, cataloging and sorting them based on semantic categories, virtually reconstructing the heritage through photogrammetric techniques and the use of certain software, and, in the case of media, communicating and spreading the word about the project, or simply broadening the network talking with peers. Such a project continually creates engagement by massively exploiting the potential of collaborative technology. In addition to managers, policy makers could take advantage of this paradigm when developing projects aimed at valorizing heritage or artworks.

Rekrei also guarantees social sustainability by showing the potential of social interaction sorted in a multiplicity of networks, as theorized by many-to-many marketing and is able to generate a fully self-sufficient organization. Rekrei is a cultural project that exemplifies the creation of a network reconciling diverse communities: academic, historical, informatics, politics as well as the broader public, in a system that enhances the characteristics of each one, through the generation of powerful and multiform resources. In this sense, the organizational sorting system that volunteers have created starting from the chat on the web platform can be compared to a self-generated and proactive human resources office.

By integrating the resources of each stakeholder in a balanced manner as proposed by the balanced centricity paradigm, Rekrei allows every stakeholder to benefit from the services generated by the project. The interaction, both human and collaborative, is based on the dynamics of relational exchange and capitalizes on the social and relational dimensions of cultural organizations.

Rekrei also represents a paradigm of heritage conservation that guarantees economic sustainability. It is based on a bottom-up orientation that derives its financial resources from the multitude of stakeholders. By offering its users the option to donate, Rekrei makes of crowdfunding an option to sustain itself. Relationships are the beating heart of cultural crowdfunding campaigns and most of the financiers of Rekrei come from the community of users that also represents the social capital of the project. The increase of the social capital is obtained through establishing relationships based on trust, emotional ties, shared values and goals representing a resource for the attainment of a goal that could not be achieved by single individuals.

5 Conclusion

This chapter has analyzed the case of Rekrei, a project developed in 2015 by two American researchers, Chance Coughenour and Matthew Vincent. The intent of this entrepreneurial project was to provide cultural heritage—that has been destroyed or dispersed due to wars, natural disasters, or other misfortunes—with new life through digital restoration. The Rekrei case is useful to understand the application of disintermediation, value co-creation and SD-L theories to a practical case. Indeed, the Rekrei project allows for the development of co-creative practice and the creation of a network of stakeholders that are both contributing and benefiting from the initiative. Rekrei represents a groundbreaking heritage conservation paradigm, which can allow for cultural, social and economic sustainability for heritage and arts-based projects. The case of Rekrei demonstrates how, to be really innovative, an entrepreneurial project needs to be strongly connected to the web, be unique by positioning itself in a different way from other similar projects, and use a universal language. In this way, a project can easily obtain a competitive advantage, thus reaching a global audience.

The study is exploratory in nature and does not mean to be exhaustive. Further research should explore the application of the aforementioned theories in other contexts and with reference to more cases.

References

1. L. Aarikka-Stenroos, E. Jaakkola, Value co-creation in knowledge intensive business services: a dyadic perspective on the joint problem solving process. Ind. Mark. Manag. **41**(1), 15–26 (2012)
2. W. Benjamin, *The Work of Art in the Age of Mechanical Reproduction* (Schocken, New York, 1936). 1960
3. Y. Benkler, Peer production, the commons, and the future of the firm. Strateg. Organ. **15**(2), 264–274 (2017)

4. E. Clemons, K.R. Lang, The decoupling of value creation from revenue: a strategic analysis of the markets for pure information goods. Inf. Technol. Manag. **4**(2–3), 259–287 (2003)
5. A. Corns, R. Shaw, High resolution 3-dimensional documentation of archaeological monuments & landscapes using airborne LiDAR. J. Cult. Herit. **10**, 72–77 (2009)
6. J. W. Creswell, J. D. Creswell, *Research design: Qualitative, quantitative, and mixed methods approaches*. Sage publications (2017)
7. K.D. Elsbach, R.I. Sutton, Acquiring organizational legitimacy through illegitimate actions: a marriage of institutional and impression management theories. Acad. Manag. J. **35**(4), 699–738 (1992)
8. M. Foucault, Governmentality, in *The Foucault Effect*, ed. by G. Burchell, C. Gordon, P. Miller (University of Chicago Press, Chicago, 1991), pp. 87–104
9. C. Grönroos, Conceptualising value co-creation: a journey to the 1970s and back to the future. J. Mark. Manag. **28**(13–14), 1520–1534 (2012)
10. G. Guidi, M. Russo, S. Ercoli, F. Remondino, A. Rizzi, F. Menna, A multi-resolution methodology for the 3D modeling of large and complex archeological areas. Int. J. Archit. Comput. **7**(1), 39–55 (2009)
11. G. Guidi, M. Russo, Reality-based and reconstructive models: digital media for cultural heritage valorization. SCIRES-IT: SCIentific RESearch Inf. Technol. **1**(2), 71–86 (2011)
12. D.B. Holt, Why do brands cause trouble? A dialectical theory of consumer culture and branding. J. Consum. Res. **29**(1), 70–90 (2002)
13. J. Howe, The rise of crowd sourcing. Wired mag. **14**(6), 1–4 (2006)
14. J. Hughes, K. Lang, If I had a song: the culture of digital community networks and its impact on the music industry. Int. J. Media Manag. **5**(3), 180–189 (2003)
15. E. Katz, Disintermediation: cutting out the middle man. *Departmental Papers (ASC)*, 162 (1988)
16. A. Lind, D. Nylén, Mapping everyday objects to digital materiality in the wheel quintet: polytempic music and participatory art. 16th International Conference on New Interfaces for Musical Expression (NIME'16), July 11–15, (Griffith University, Brisbane, Australia, 2016), pp. 84–89
17. R.F. Lusch, F.E. Webster Jr., A stakeholder-unifying, cocreation philosophy for marketing. J. Macromark. **31**(2), 129–134 (2011)
18. J.W. Meyer, B. Rowan, Institutionalized organizations: formal structure as myth and ceremony. Am. J. Sociol. **26**(1), 340–363 (1977)
19. B. Newman, Inventing the future of the arts: seven digital trends that present challenges and opportunities for success in the cultural sector, in *20under40: Reinventing the Arts and Arts Education for the 21st Century*, ed. by E. Clapp (Authorhouse, Bloomington, IN, 2010), pp. 3–19
20. C.K. Prahalad, V. Ramaswamy, *The Future of Competition: Co-Creating Unique Value with Customers* (Harvard Business School Press, Boston, 2004)
21. V. Ramaswamy, K. Ozcan, What is co-creation? An interactional creation framework and its implications for value creation. J. Bus. Res. **84**, 196–205 (2018)
22. F. Remondino, Heritage recording and 3D modeling with photogrammetry and 3D scanning. Remote Sens. **3**(6), 1104–1138 (2011)
23. N.S. Rose, *Powers of Freedom: Reframing Political Thought* (Cambridge University Press, Cambridge, 1999)
24. A. Styliadis, I.I. Akbaylar, D.A. Papadopoulou, N.D. Hasanagas, S.A. Roussa, L. Sexidis, Metadata-based heritage sites modeling with e-learning functionality. J. Cult. Herita. **10**(2), 296–312 (2009)
25. S.L. Vargo, R.F. Lusch, Evolving to a new dominant logic for marketing. J. Mark. **68**(1), 1–17 (2004)
26. P.P.M. Vargo, M.A. Akaka, On value and value co-creation: a service systems and service logic perspective. Eur. Manag. J. **26**(3), 145–152 (2008)
27. D. Zwick, S.K. Bonsu, A. Darmody, Putting consumers to work: 'co-creation' and new marketing govern-mentality. J. Consum. Cult. **8**(2), 163–96 (2008)

Game-Based Learning Research at the Danube University Krems' Center for Applied Game Studies

Nikolaus König, Alexander Pfeiffer, Thomas Wernbacher and Natalie Denk

Abstract For more than 13 years, the Center for Applied Game Studies at the Danube University Krems, Austria, has conducted game-based learning research. Guided by the pragmatic aim to enable classroom integration of game-based learning on a broad scale, this research has centered on the development of a gbl-toolkit for (and in cooperation with) teachers, enabling them to use game-based learning strategies within the context of current educational practice, while at the same time expanding their own knowledge and skills as gbl-experts. Currently, the toolkit contains tools to evaluate the pedagogical potential of digital games, to plan and conduct gbl-projects, and to exchange with other teachers and form an active gbl-community. Next steps will be an educational game design tool and the first in a series of educational game editors for teachers. This chapter will present these tools and discuss lessons learned from their development.

1 Overview

Since its foundation in 2006, the Danube University Krems' Center for Applied Game Studies has examined, argued for, and put to the test the pedagogical potential of digital gaming media and their value for educational practice.

One of the center's earliest Master courses (titled "Media- and Game Pedagogy" or—due to the Austrian weakness for compound-nouns—"MedienSpielPädagogik")

N. König · A. Pfeiffer (✉) · T. Wernbacher · N. Denk
Danube University Krems, Krems, Austria
e-mail: alexander.pfeiffer@donau-uni.ac.at

N. König
e-mail: nikolaus.koenig@donau-uni.ac.at

T. Wernbacher
e-mail: thomas.wernbacher@donau-uni.ac.at

N. Denk
e-mail: natalie.denk@donau-uni.ac.at

D. Seychell and A. Dingli (eds.), *Rediscovering Heritage Through Technology*, Studies in Computational Intelligence 859,
https://doi.org/10.1007/978-3-030-36107-5_7

was one of the first academic training opportunities worldwide centered on the theory and practice of game-based learning and (with regular updates to its curriculum) persists until today.

At the same time, some of the Center's major research projects have been focused on the conditions under which game-based learning can be successful in educational practice, guided by the idea that the potential of digital games to enable meaningful learning experiences can only unfold in an adequate educational environment, and is dependent on skilled teachers and their ability to facilitate connected learning experiences through digital games. This research has laid the foundation for the Center's teaching activities, but at the same time, the involvement of its students in these research projects has greatly contributed to the Center's research profile. In the following, an overview will be given of the Center's major game-based learning related research projects and their relation to educational practice in Austria, with an emphasis on the ongoing effort of developing a game-based learning toolkit for teachers.

2 Background: Didactic Scenarios of Digital Game-Based Learning (2007–2010)

Starting in 2007, the Center's first major research project in the field of digital-game based learning was focused using digital games to close the gap between school life and the daily (media) experiences of children and youths—and by finding out how digital games might be used in the classroom in order to achieve this. At the center of this investigation were questions regarding the feasibility of game-based learning approaches in educational practice, especially within the Austrian school system (and even more specifically, within the austrian-wide educational concept of the new middle school "Neue Mittelschule" NMS).

As the aim of this project—the first in a series of investigations funded by the Austrian Ministry of Education—was to harvest the potential of digital games to close the gap between school life and student's daily (media) experiences, the project used a series of commercial off-the-shelf games that enjoyed a certain popularity among students; on the other hand, most of these games had been used in the educational context before, so it would be possible to distinguish whether specific problems were caused by the game products themselves, or by the ways they were applied in the context of the project. Over the course of the project, Austrian teachers worked closely with experts from the fields of game studies and media pedagogy to (1) develop a basic understanding of the medium-specific properties of digital games and their educational potential, (2) develop didactic concepts for using specific games in their own classrooms, and to make their own decisions regarding game choice and didactic goals, and (3) apply the concepts in their classroom, reflect on their success, and involve their students in a discussion of the pros and cons of digital game-based learning, based on the shared experience of using a game in their classroom.

The project was accompanied by a qualitative as well as quantitative evaluation as a basis for future strategic decisions regarding the use of digital game-based learning in Austrian schools.

Not only did the project establish the pragmatic perspective that would determine the Center's research activities until today; it also led to the formulation of 12 "Theses regarding the use of computer games in the classroom" [7].

3 The Game-Based Learning Toolkit (2013/14)

At the late 2000s and the beginning of the 2010s, there already was wide academic consensus that digital games have the potential to enable meaningful learning experiences (e.g. [1, 3–5, 8, 11]). But attempts to actually use game-based learning in educational practice were still rare, and if they were made [9], they were designed as more or less radical alternatives to traditional education and, even more importantly, accessible only to a selected few.

In Austria, there was high demand for game-based learning strategies due to the reorganization of the school curricula and the introduction of the educational concept of the new middle school (NMS) with its core principles of Differentiation, Individualization and Personalization. However, the needs of the New Middle School could not be met with exceptional pilot-projects alone but required strategies to consolidate game-based learning approaches with existing educational practice, and to facilitate the best possible application of game-based learning within the restrictions of the (Austrian) educational system; the goal was enabling the use of digital games as catalysts for connected learning experiences not as an alternative to, but as a part of the existing school model. A second project ("Gaming Media and their Application in Educational Practice") was therefore initiated (once again funded by the Austrian Ministry of Education).

Building on the earlier project's theses as well as the demands of the Austrian educational system, the project's aims was to consolidate game-based learning with contemporary models of educational practice, while at the same time facilitating the building of teachers' competencies on a broad scale. The goal of the project was to cooperate with game experts as well as educators in the development of a series of tools that teachers could use to:

1. identify the educational potential of any given digital game, while at the same time developing their own understanding of the medium—specific properties of digital games (the "Analysis Tool");
2. develop suitable didactic scenarios to use these games in their educational practice, accounting for the specific needs of their respective students as well as the possibilities and obstacles at their own school location (the "Application Tool");
3. form and participate in a game-based learning community, while at the same time game suggestions, discussing game-based learning scenarios and assist each other with their projects (the Online Platform);

4. develop their own game projects using an easy to use game-editor, systematically translating their pedagogical and didactical needs into game design choices (The "Game Design Tool and Editor").

The guiding idea behind these tools' development was that the integration of game-based learning into educational practice must be addressed in a sufficiently pragmatic manner: game-based learning cannot be regarded as a fixed procedure which can be imposed on any educational setting alike but is a contingent set of principles which need to be adapted for every particular school, classroom and even student. While rooted in sound educational and game theoretical concepts, the iterative development process involving experts from both areas was aimed at creating tools that were flexible enough to be adapted by teachers for their own student's needs and classroom situations.

3.1 The Analysis Tool

For the development of the analysis tool, game-experts (graduates in game studies and professionals in various game-related fields) as well as experts in education (teachers holding a master's degree in media- and game-pedagogy) were put together in teams. These teams were presented with a preliminary questionnaire based on a structured translation of game-theoretical concepts into pedagogic and didactic principles, complete with detailed comments on the concepts and principles used (this outline would later be adapted, refined, streamlined and reformulated according to the teams' findings in the subsequent project steps). Additionally, each team picked a specific game, which would later serve as a testing ground for the development and refinement of the "analysis-tool". In the following weeks each of the teams assessed the preliminary questionnaire, based on their professional experience and the analyses of the example games in order to assess what adaptations they deemed necessary to help teachers (1) identify a game's pedagogical potential and (2) gain an understanding of the distinct qualities of the medium. Based on the thorough feedback of the teams, the questionnaire and comments were then restructured, adapted and refined.

The final version of this "analysis tool" consists of six basic categories which are further divided into sub-categories to enable a more detailed insight into the respective category. The questions constituting these categories/sub-categories are separated in general questions, which highlight the most important aspects of the respective category, and advanced questions, which either help to examine a category more closely, or help to get a better grasp on those categories which are more difficult to use.

The first category—General Information—assesses the game's basic production data, gaming platform and operating system. It is asked whether the game has been designed for educational or entertainment purposes, or whether it is a game-related software product rather than an actual game. While this first category is very general, it encourages to perceive games as embedded in a broader technological, economic and creative context. The second category—Form and Technology (Game Product) includes technological requirements, the game's purported theme and its presentation.

All in all, this category presents a soft entry to the analysis by restricting itself to aspects which can easily be assessed, even before the game is actually played. The third category—Content (Game World)—requires a closer look into the game, but deals with aspects that computer games share with more traditional media like film or literature: what is the setting of the game, what are the data and facts it contains, and what kinds of narratives (if any) does it rely on?

The remaining three categories finally take the specific qualities of digital games into account. These game-specific categories of the "analysis-tool" are closely related to the MDA framework for computer game analysis [6].

The fourth category—Mechanics (Game System)—examines the game's core mechanics as well as the flexibility of the game system (room for errors/alternative solutions), as well as the relation between game system and learning experiences.

The fifth category—dynamics (player actions)—deals with game verbs and complexity (player actions and depth of cognitive involvement, see [12]), with matters of roles and identity and with player interactions (cooperation vs. competition). The aim of the sixth category—Aesthetics (player experience)—is to make the volatile field of gameplay/learning experience tangible by highlighting possible fields of experience, asking teachers to reflect on their own gameplay experiences by encouraging the observation and discussion of other players' gameplay experiences.

As is obvious in the description of the various categories, the goal of the "analysis-tool" is not to conduct a thorough and authoritative analysis of a game, but to guide teachers in approaching digital games from a sound perspective, getting a grasp on their medium-specific qualities while at the same time developing the skills necessary to assess the pedagogical potential of a specific game on various levels of the medium.

3.2 *The Application Tool*

The goal of the second project phase was to design a tool which assists teachers to apply the pedagogical potential of a specific game to their own classroom by enabling the design of classroom-specific game-based learning scenarios. In this phase, it was crucial to reconcile the pedagogical potential of the medium with the strict limitations of educational reality. The teams were expanded by teachers who had no prior experience with digital games to ensure that the tools were accessible even without a prior understanding of game-based learning principles.

The "application-tool" was designed to be used as a flexible supporting tool for the development and application of game-based learning scenarios in the teachers' own educational setting. As a result of workshop discussion and subsequent testing of the tool during the pilot projects, four key areas have proven relevant for a pragmatic, yet meaningful implementation of game-based learning strategies in educational practice: (1) the educational objectives which can be pursued based on the game; (2) the added value and limitations of employing a game-based learning strategy to reach these objectives, (3) the methods required to tap the pedagogical potential of the game in the classroom and (4) the problems that must be solved before game-based learning strategies can be successfully employed in a specific educational setting.

1. Educational objectives: the definition of educational objectives forms the starting point of meaningful learning design [13]. What are the core questions which define how these concepts are approached? What are common misunderstandings in regard to these concepts, and how are they addressed by the game? And finally, how can the game help to enable meaningful learning and understanding [10]?
2. Added value and limitations: game-based learning is not an end in itself. What can the game add to a meaningful learning process? Can the objectives be reached in alternative ways, and what are the upsides and downsides of these approaches? Which aspects of the educational goals are not sufficiently addressed by the game, and what complementary strategies are required?
3. Methods: the success of game-based learning is dependent on how they are integrated in the learning situation: what preparations are necessary for a specific game-based learning project, and how must the gameplay be hosted to be meaningful and effective? How does the transition from playing a game back to the educational setting take place, and what will students need in order to benefit from the game?
4. Problems: the integration of a new medium into educational practice is not without problems. It is important to be aware of these problems, in order to counter or minimize them early enough. Will expert help be needed, or is it necessary to adapt the scenario in order to minimize these problems? Is it possible to conduct the game-based learning project within the class schedule? Can the schedule be adjusted, or does the project itself need to be adapted? What support will students need to comprehend the game's topic and achieve the educational goals? Are colleagues, parents or even student skeptical towards the use of computer games in the classroom? How can this skepticism be reduced in order to create a cooperative educational environment?

3.3 The Online Platform

In the final phase of the project, the basic demands of the Online Platform (whose actual development would take place in another project, see Chap. 3) were defined. While it was clear that the Online Platform would also provide access to the previously developed tools, its main function was to provide adaptable entry points depending on the user's expertise and previous experience with computer games and game-based learning strategies. The basic goal was to enable dynamic and discursive competence-building within the teaching community by fostering a "choose what to use" approach; according to their respective demands and competencies, teachers were to be able to choose different ways to use the platform (for details, see Sect. 4).

4 The First Application: Game-Based Learning to Alleviate Early School Leaving in Malta (EU)

Contrary to the first two tools (The "Analysis Tool" and the "Application Tool"), which were developed, tested and refined in the context of the Austrian education system, the development of the Online Platform was conducted as part of an Erasmus Plus Strategic Partnership between five partners: the Malta Information Technology Agency (MITA), the University of Malta, St. Margaret's College Malta, Nystromska School Sweden, and Danube University Krems, Austria. While the development of the online tool was only one segment of this partnership, whose main goal was to explore the potential of game-based learning to alleviate early school leaving, this embedded approach enabled to develop and test the platform on a broader scale than a national project would have allowed to, while at the same time providing a first application for the previously developed tools.

As it was one of the main aims of the project to enable teachers to become "agents of change" within their schools by expanding, practicing and sharing their expertise in game-based learning, the platform needed to provide entry points for teachers of all skill levels regarding game-based learning, ranging from game-based learning experts to novices and even sceptics [2]. Based on the previous project's output, the following ways to engage with the game-based learning community via the platform become the guiding principles of its development and refinement:

1. Scuttle: the platform contains a database with existing game-based learning scenarios and game analyses, as well as commented documentations from game-based learning projects that have already been used in the classroom. Furthermore, the platform contains background material on digital games and game-based learning. Looking through the material on the platform, teachers are able to gain first insights into the use of computer games in educational practice and get familiar with the work of colleagues and the theoretical foundations of game-based learning.
2. Dabble: it is not necessary to have prior experience with game-based learning to engage in the game-based learning discourse. In the discussion forums, teachers without prior knowledge about game-based learning can be assisted by more experienced colleagues, but they can also contribute insights from their own teaching experience. The discussion forums help teachers get familiar with the topic and—most importantly—become part of the game-based learning community.
3. Sample: while the design of the "analysis tool" and "application tool" aims to enable classroom-specific game-based learning scenarios, it is possible and valid to draw on existing scenarios and use them in one's own classroom. While these scenarios have not been adapted to the specific educational setting, they will still

provide an opportunity to gain first hand experiences with game-based learning and highlight the demands and opportunities of the specific classroom, thereby inspiring the development of more specific scenarios.

4. Grapple: teachers feeling sufficiently familiar with the basic principles of game-based learning can decide to use the "analysis tool" to identify the pedagogical potential of a particular game. The tool may lay the foundation for designing a game-based learning scenario, or it may simply be used to gain inside into a game that the teacher's students keep bringing up in the classroom. In any case, using the analysis tool is the first step of developing an autonomous approach towards game-based learning.
5. Embrace: as soon as teachers feel ready, they can develop their own game-based learning scenarios using the "application tool". In addition to using the tool, teachers can discuss problems and ideas with colleagues in the discussion forums, or in the comments section of a project they have shared on the platform.
6. Lead: using the platform, teachers increasingly gain experience and confidence as game-based learning experts. As their own competencies continue to grow, they might choose to actively engage in supporting their colleagues in getting a hold of game-based learning and its application to the classroom. As "teachers leaders", they become an integral part of the game-based learning community, helping others in becoming familiar with the pedagogical potential of digital games and their application in educational practice.

5 Outlook

The next tools in the game-based learning toolkit will be an Educational Game Design Tool as well as an Educational Game Editor (EGE-1), which enables teachers to design and create their own games for/with their students. The development of these tools will once again be embedded in an Erasmus Plus Strategic Partnership, in Cooperation with the Malta Information Technology Agency (MITA), the University of Luxembourg, the Science Centre Malta, the game development company waza! UG, Germany, and the Research and Innovation Management GmbH, Austria.

The Game Design Tool will be based on the research conducted for the development of the "Analysis Tool". However, the creation of the new tool will require a basic re-design of the questionnaire and its structure, as well as the comments on the underlying concepts and principles. The goal behind this tools' development is to provide easily accessible guidelines for teachers without a background in game-design and -development that will enable them to:

1. assess whether specific educational goals can benefit from a game-based learning approach using a self-made game;
2. define a realistic scope for the project and translate the educational topic into a game concept;

3. decide which game form best support the educational goals and choose an appropriate game editor;
4. implement their game concept step-by-step using a game editor.

Mirroring the insights from the creation of the analysis tool, the game design tool will translate game-design principles into educational concepts to enable teachers to base game-design decisions on their own skill-set as educators. (However, insights from this development process will be used to assess how later adaptations for a similar tool might look like that allows game designers to base the pedagogic/didactic decisions in the development of educational games on their existing expertise in game-design).

The Game Editor (EGE-1) will be the first in a series of low-threshold educational game-editors. Each of these editors will highlight another primary mechanic, so that teachers have appropriate editors available depending on the choices made using the Educational Game Design Tool. EGE-1 will use a dual-decision-tree mechanic that can make action-consequence relations tangible by means of simple interactive narration. These editor's will be developed in cooperation with teachers and education experts, the goal being to employ interface conventions, design strategies and vocabulary stemming from the educational context in order to enable easy translation of educational aims into game-design decisions.

Toolkit links:

- TOOLKIT Plattform: http://toolkit-gbl.com
- Register for free: https://toolkit-gbl.com/register
- Further information regarding the project team: https://www.mita.gov.mt/en/DigitalOutReach/gamebasedlearning/Pages/Game-Based-Learning.aspx
- Guidebook "Gamify Your Classroom!": https://toolkit-gbl.com/downloads/guidebook.

References

1. L.A. Annetta, The "i's" have it: a framework for serious educational game design. Rev. General Psychol. **14**(2), 105–113 (2010)
2. N. Denk, Einblicke in die toolkit plattform. Medienimpulse **55**(2) (2017)
3. A. Dingli, C. Bonello, G. Farrugia, Games in the classroom, in *ICERI2013 Proceedings* (IATED, 2013), pp. 2846–2852
4. J. Fromme, B. Jörissen, A. Unger, Bildungspotenziale digitaler spiele und spielkulturen. MedienPädagogik: Zeitschrift für Theorie und Praxis der Medienbildung **15**, 1–23 (2008)
5. J.P. Gee, Video games, learning, and "content", in *Games: Purpose and Potential in Education* (Springer, Berlin, 2009), pp. 43–53
6. R. Hunicke, M. LeBlanc, R. Zubek, Mda: a formal approach to game design and game research, in *Proceedings of the AAAI Workshop on Challenges in Game AI*, vol. 4 (2004), p. 1722
7. K. Mitgutsch, M. Wagner, Didaktische szenarien des digital game based learning. Krems/Wien (2008)
8. M. Pivec, M. Moretti, *Game Based Learning: Discover the Pleasure of Learning* (Pabst Science Publishing, 2008)

9. K. Salen Tekinbaş, R. Torres, L. Wolozin, R. Rufo-Tepper, A. Shapiro, *Quest to Learn: Developing the School for Digital Kids* (MIT Press, New York, 2010)
10. M. Schratz, B. Weiser, Dimensionen für die entwicklung der qualität von unterricht. J. für schulentwicklung **6**(4), 36–47 (2002)
11. M. Wagner, S. Gabriel, Didaktische szenarien des digital game based learning. *Handreichung für Lehrkräfte, Wien: bm: ukk* (letzter Zugriff: 20.06. 2017) (2011). http://pro.kphvie.ac.at/fileadmin/pro/pro/gbl/unterlagen/handreichung_dgbl_mai.pdf
12. N.L. Webb, Depth-of-knowledge levels for four content areas. Lang. Arts **28**(March) (2002)
13. T. Westfall-Greiter, C. Hofbauer, Shared leadership setzt teacher leaders voraus: Lerndesigner/innen im feld der neuen mittelschule. J. für Schulentwicklung **4**, 8–14 (2010)

Virtual Reality

Virtual reality is like dreaming with your eyes open.
Brennan Spiegel

A Case Study into the User Experience of an Application of Virtual Reality at the Saint Paul's Catacombs, Malta

Jeremy Grech, Mark Bugeja and Dylan Seychell

Abstract Several historic sites in the Maltese Islands are not easy to explore; their construction typology or fragile state of conservation can impede accessibility to all. The Saint Paul's Catacombs in Rabat is one such historic site. With this in mind, can Virtual Reality (VR) be used to add value to the user experience (UX) of someone exploring these heritage sites? And if so, how can this value be added? This chapter attempts to answer these questions through a qualitative study of a VR system created for the St Paul's Catacombs in Malta. An assessment of similar applications created for other cultural heritage sites around the world is conducted. This chapter also describes how this VR experience was created through the use of Game Engines, Laser Scanning and other technologies.

1 Introduction

Virtual Reality (VR) technology has allowed users to explore virtual worlds in many different contexts, from Game worlds to simulations to immersive fictional worlds for storytelling. It stands to reason that VR has also been used within the context of cultural heritage, albeit implemented in different ways.

Mark Bugeja is now a research support officer at the University of Malta, but did this work earlier as a lecturer of Saint Martins Institute of Higher Education.
Dylan Seychell is now a lecturer at the University of Malta, but did this work earlier as a lecturer of Saint Martins Institute of Higher Education.

J. Grech · D. Seychell
Saint Martin's Institute of Higher Education, Hamrun, Malta
e-mail: jgrech@stmartins.edu

D. Seychell
e-mail: dylan.seychell@ieee.org

M. Bugeja (✉)
University of Malta, Msida, Malta
e-mail: mark.bugeja@um.edu.mt

D. Seychell and A. Dingli (eds.), *Rediscovering Heritage Through Technology*, Studies in Computational Intelligence 859,
https://doi.org/10.1007/978-3-030-36107-5_8

In this chapter, one would first find an argument identifying a problem that exists when one explores some cultural heritage sites. In particular, the chapter will be discussing issues that one can find when visiting heritage sites which have hard to reach areas or areas with restricted access. The lack of access can be down to one's mobility issues (for example, people in wheelchairs) or because specific locations within the heritage site are too fragile. Hence, it can be awkward or even dangerous to allow visitor access. The argument made here is that these areas (or locations) might still be of cultural and historical significance. If one cannot access them, then it means that it is more challenging to communicate and share the importance of the site to visitors.

In some cases, it is not possible to provide physical access to these sites as doing so will endanger or damage the site. VR is one way in which this problem can be overcome. The chapter then progresses to describe the motivations behind the creation of the Virtual Reality Experience of the Saint Paul's Catacombs in Rabat, Malta. The Saint Paul's Catacomb a Maltese heritage site which does provide several accessibility challenges. Having said this, one might then ask, with the implementation of this VR experience, how has the user experience changed? In other words; Can this implementation of the catacombs environment in virtual reality add value to the user experience when one is exploring said heritage site? And if so, how can this value be added?

A literature review is conducted to explain how VR experiences have been used or implemented in (or about) other cultural heritage sites around the world. Their implementations and methodologies are critically analysed, especially their implications in relation to the user experience these VR implementations provided. The chapter then moves on to describe how the Saint Paul's Virtual Reality experience was developed. This experience was designed to:

- Make it easier and more intuitive to explore laser scans made of the catacombs
- Provide means by which a person with limited mobility would be able to explore said catacombs
- Make the Virtual Reality experience as realistic as possible and
- Create an experience which would immerse the user as much as possible.

After that, one would find a comparison of the user experience one can have when exploring Saint Paul's Catacombs with and without the VR experience. It is then shown that said VR experience does add value to the overall user experience that one can have in visiting this site. This is because it removes barriers, allows access to hard to reach areas and also allows one to view the site with more (original) detail than one can physically. It must be said that this VR experience cannot and should not replace the real world experience. Instead, it should be used to enhance the physical experience itself.

The final section of this chapter concludes the argument by stating the results obtained. These results were obtained through the evaluation conducted in the previous section describing how the goals of the VR implementation were met and how the research questions made were answered. An analysis of what future work that could be carried out to enhance the shortcomings of this chapter is also included here.

2 Identification of Problem

Heritage sites are places in which historic buildings and remains are culturally and historically significant. It is, therefore, important that one preserves these sites without altering them as much as possible. It is the case that some locations within heritage sights are in areas which might be very hard to reach, or even dangerous to access. Moreover, some heritage sites are also incredibly delicate and sensitive to several different factors, such as humidity, temperature, air quality, light etc., which can be introduced by visitors, unwittingly or otherwise, to the site.

Altering such sites also presents problems. In most cases, changing the structure of such a historical site is incredibly frowned upon, logistically and technically challenging, expensive. Even if any alterations were to be made, they should be done with great care as conservation and preservation would be of the utmost importance. For these reasons, this means that, sometimes, it is not possible to make a heritage site more accessible, as in doing so, one might damage the site itself.

Physical accessibility is, therefore, a genuine problem, especially when one wants to open up such sites to be enjoyed by the general public. It is important to note that this problem is not only limited to people who have mobility issues. In his Doctoral thesis, when talking about the physical access to several Maltese heritage sites, Glen Farrugia wrote:

> Archaeological sites are often made up of intricate and complicated layouts [12]. This aspect offered and still presents, a great challenge to curators in preparing heritage spaces for visiting. A case in point is the type of physical environment which characterises the Maltese Late Roman catacombs (and other catacombs located elsewhere). These sites are made up of steep rock-cut uneven steps leading to narrow corridors and chambers hewed out of the living rock. The only way of making the catacombs fully accessible is by altering the original architectural fabric of these sites, something which in the 1980s was already considered as unethical. Therefore, site visits at the catacombs and similar archaeological sites were limited to persons who were fit to do so. (p. 91) [10]

A possible solution to overcome this problem of accessibility is to use Virtual Reality technology within these heritage sites. Hence one can call it Virtual Cultural Heritage. Cameron and Kenderdine write:

> The term 'virtual cultural heritage' is generally accepted to mean: virtual reality (3D and 4D computational and computer graphics systems that support real-time, immersive, and interactive operations), employed specifically for the presentation, preservation, conservation, and documentation of natural and cultural heritage. (p. 10) [5]

3 Motivation

The Saint Paul's Catacombs, located in Rabat Malta was constructed during the Roman occupation of the island and were used as cemeteries during the 7th and 8th Centuries AD. These catacombs were carved out of the limestone, thus creating a complex of interconnected underground tombs which cover well over 2000 square meters [14].

The site where these catacombs are found consists of two large areas known as the Saint Paul's and Saint Agatha's Catacombs. Evidence shows that early Christians used these catacombs in Malta. The catacombs were largely abandoned during the Saracenic period, but some of the catacombs were reused again in the 13th Century AD when Malta was re-Christianised. Eventually, the catacombs were again abandoned and fell into disrepair. It was only in 1984 when A. A. Caruana cleared the debris and all passages [14].

At the time of writing, the site is managed and maintained by Heritage Malta a national agency for museums, conservation practice and cultural heritage created by the Cultural Heritage act enacted in 2002 [15]. It is worth mentioning that the Saint Paul's Catacombs are currently on a tentative UNESCO World Heritage Site list [24]. In 2008/09, Heritage Malta started an ERDF (European Regional Development Fund) part funded project in which 3D documentation of some of the Saint Paul's catacombs was to be carried out [16].

In the project mentioned above, 3D laser scans were made by surveying several catacombs located at the site. In March 2013, Heritage Malta had signed a memorandum of understanding with Saint Martin's Institute of Higher Education (known back then as Saint Martin's Institute of Information Technology) such that:

> the two entities collaborating closely on various projects in an attempt to pursue the mission undertaken by Heritage Malta to modernise the museums and sites entrusted to the agency making them more accessible and an experience worth remembering for visitors [17]

As a result of this understanding, Saint Martin's was allowed to develop a Virtual Reality (VR) Experience of the Saint Paul's Catacombs. This experience was created using the 3D models obtained from the laser scans made from the above mentioned ERDF project. The motivation behind the VR experience to be developed were:

1. To create an application which made it easy to explore the 3D models generated without having to use 3D model viewing applications. This goal was requested due to several reasons. Chief amongst them were the following:
    - The orientation of specific models might be different from that of the 3D model viewing software. Thus it would mean that before anyone could start viewing the model, one would need to orient it correctly and then try to navigate through the model.
    - When viewing a 3D model in such a software system, one would need to use the pan, rotate and zoom functionality to navigate the model. This means that navigation is not as easy as it can be, and it might take a considerable amount of practice to master.

- Most 3D viewing software don't consider camera collision. This means that the user can pass the virtual camera, i.e. the point of view from which the user is exploring the model, through walls within the model. Thus, breaking any immersion achieved and also contributing to a sense of disorientation, especially given the labyrinth nature of the catacombs themselves.
- In general, 3D model viewing software does not make use of real-time realistic shading. This is because most of these applications usually only feature a relatively simplified preview of the item to be rendered due to performance reasons. If one decides to do so, one would be able to render that particular view of the model using more sophisticated (and probably more computationally expensive) shading techniques. This difference can be shown in Fig. 1.

2. To allow one to explore and experience the catacombs for themselves, even if one has limited mobility. The Saint Paul's Catacombs, having been built in around the 7th and 8th centuries and been used as cemeteries [14], were not built with ease of accessibility in mind. A VR experience would be ideal for helping said visitors to at least get a taste of what it would be like to be in catacombs without being physically present. This is especially true when one considers that the catacombs themselves cannot be altered to accommodate ramps, lifts and other modes of access due to their construction. Therefore a VR experience would improve accessibility to this historical site which attracts around 80,000 people annually [7].
3. To allow visitors and researchers to be immersed in a high fidelity environment as well as to provide one with the possibility to learn new things about the site itself in safety. This safety not only extends to the users of this VR system but also to the historical site itself. VR allows one to be immersed in the environment shown. According to Biocca and Levy:

   > The blocking of sensory impressions from physical reality is a crucial part of the most compelling VR experiences. The senses are immersed in the virtual world; the body is entrusted to a reality engine (p. 135) [2]

   Thus, it makes sense that by developing a high fidelity experience, it would allow one to be immersed.

Which brings us to the question: Can an implementation the Saint Paul's Catacombs environment in virtual reality add value to the user experience when one is exploring said heritage site? And if so, how can this value be added? These questions are going to be tackled in Sect. 6.

4 Literature Review

The topic of Virtual Reality is fast becoming one of the more popular fields of study. This is down to the fact the Virtual Reality devices have recently become more available and also affordable. The availability of such tools allows more people

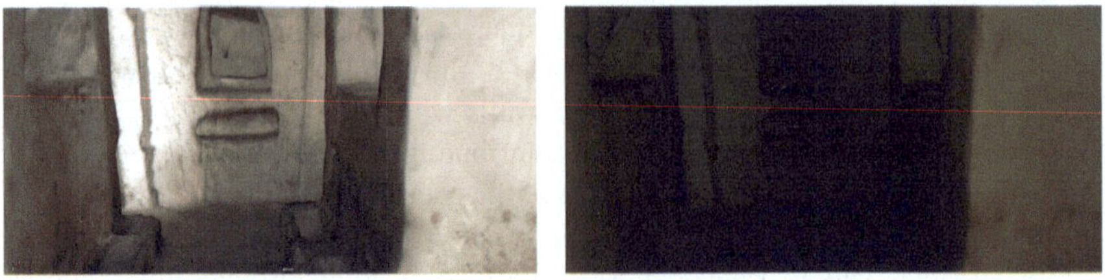

(a) Maya's Scene View (b) Maya's Arnold renderer

Fig. 1 A scene from Catacomb dubbed no. 14 showing the difference between Maya's Scene View and Maya's Arnold Renderer. *Source* Author

to experiment with them, develop new experiences with them and also study these experiences. Virtual reality has been used and studied in several different fields from gaming and game development, to education, to the study of disorders as well as Cultural Heritage.

The use of Virtual Reality within Cultural Heritage is known as **Virtual Cultural Heritage** [5] and it has been in use in various implementations around the world. One can imagine that the use of such technology can enhance the way people experience cultural heritage.

4.1 How Virtual Reality Is Currently Used in Cultural Heritage

Bekele et al. conducted a survey of the use of Augmented, Virtual and Mixed Reality for Cultural Heritage. According to them

> The majority of VR applications serve virtual museums, followed by education, reconstruction and exploration purposes in that order. [1]

They go on to state that most of the applications developed tended to be semi-immersive. What they meant was that most of the time, users are presented with a pre-rendered virtual scene and view it through the use of 3D stereo displays. Fully immersive systems are not as widespread at the time of writing. However, they do support exploration themes. Reference [1] As one can see later on (in Sect. 5), a fully immersive Virtual Reality system has been opted for and developed for Saint Paul's Catacombs to help users get more access to and explore Saint Paul's Catacombs.

4.1.1 The Use of Artist-Created 3D Models for VR Experiences of Heritage Sites

Choirokoitia–Cyprus

The more traditional way in which one can obtain a 3D model of a heritage site is through the use of 3D modelling. This is done by taking a lot of reference images and then having a 3D artist reconstruct the site through the use of a 3D modelling software. This was the exact approach taken by Chirstofi et al. when creating a VR tour for the archaeological site of Choirokoitia, Cyprus. What is interesting to note is the results obtained from the experiments run of the VR system. In this paper, they tested participants on:

- **Presence**—the feeling that the participant was actually there within the virtual site
- **Learning**—the amount of information that they had learned in their exploration of the virtual site (through the means of a test) and
- **Interest**—through the means of a questionnaire where the participant's interest in Cultural heritage was measured before and after their virtual visit in Choirokoitia.

When the results were analysed, it was noticed that the presence of the participants scored a median value of 3.5 out of 5. The median score of the knowledge learnt before the experience was that of 4.11 out of 10, with a standard deviation of 1.37. The median score of the knowledge learnt after the VR experience changed to a median score of 6.84 out of 10, with a standard deviation of 2.035. Upon comparison, these results indicated that the VR experience had helped participants to learn more about the site. Finally, it was found that the interest of the participants did not change after they had visited the virtual site [8]. This example shows that Virtual Reality can help users learn more about cultural heritage.

It is the author's opinion that if Virtual Reality is used in conjunction with traditional learning that one can obtain from visiting heritage sites, then the learning of the users can be enhanced even more. It is also in the opinion of the author that a more realistic Virtual Reality Experience would, in turn, make the user's presence within the experience to be significantly improved.

Casal de Freiria, Cascais–Portugal

Another example of the use of CAD (Computer Aided Design) applications to create a Virtual Reality experience can be found with the work described by Helena Rua and Pedro Alvito. In their article, one can read about how such CAD applications were used to reconstruct the Roman villa of Casal de Freiria in Cascais, Portugal [21]. Rua and Alvito also describe how game engines can be used to aid scientific visualisation [21]. Such visualisation is one of the goals for the VR experience proposed. This work shows that it is possible to create highly detailed models of archaeological sites which could also be used as tools for the 'interactive study of the unwritten past' [21] (also known as **Virtual Archaeology**). It has to be said that these projects (i.e. those described by Rua and Alvito and by Christofi et al.) were used in applications where the historical sites in question were relatively easier to model as they were based on ruins of buildings that are found above ground.

Virtual Archaeology is a scientific activity that is sustainable and complies with the UNESCO Charter of the Preservation of Digital Heritage [19]. Given that the Saint Paul's Catacombs are on UNESCO's tentative list for becoming a World Heritage site [24], it follows that having a Virtual Reality tool, such as that proposed by this chapter, would in turn aid Heritage Malta (the current curators of the Saint Paul's Catacombs) to preserve the site as well as study it.

4.1.2 Laser Scanning Projects for Capturing Highly Accurate 3D Models of Cultural Heritage Sites

3D Scanning of the Herculaneum, Naples–Italy

In 2001, an agreement between *Packard Humanities Institute* and *Soprintendenza Archeological di Pompei*, lead to the creation of the Herculaneum Conservation Project. A highly detailed model of the interior and exterior of the Herculaneum's Suburban Baths was created through the use of a laser scanner (DeltaSphere-300) as well as theodolites and photogrammetry. This model was created in sections and then stitched together to form the complete model. The model is very accurate. All of the points within the obtained 3D model have an accuracy of 3 mm. The final textured 3D model created is then fed to a 3D graphics software program where the model can be viewed and analysed [3].

Such a project managed to create a highly detailed 3D model. As mentioned by Barizi et al., this model can then be used in several different ways. The model created can be used by conservationists, archaeologists and technical experts. It could also be made accessible to the public [3]. However, it was not mentioned how such a model could be made available to the public. It is in the author's opinion that the use of such a model would be an ideal candidate for the creation of a fully immersive 3D experience that one would be able to explore. Such an approach was similarly taken for the Saint Paul's Catacombs as later described in Sect. 5.

From 3D Laser Scanning to VR Visualisation

1. **The Rovina Paradigm**
   In their paper, Calisi et al. propose the Rovina Paradigm to improve:

   > the state of the art in measuring, documentation and classification (and thus indirectly supporting diagnosis activities) through a novel approach to surveying, data management and fruition based on three main components:
   > • **DigiRo**, an automated robot for collecting data with high-precision sensors, including laser scanners and cameras;
   > • **ARIS**, the cloud-based Archaeological Information System, to manage, share and elaborate data in the form of photo-realistic and metrically precise 3D models of the explored sites;
   > • **Web and VR Visualizers**, that allow to virtually navigate the 3D models through a very intuitive interface which also allows for an immersive experience. [4]

   This project has already been used to scan and visualise the Roman Catacombs of Priscilla in Rome, Italy. The virtual reality environment created was then showcased at the Digital Heritage EXPO in 2015 and Maker Faire in Rome 2015 [4]. The level of detail used within the Virtual Reality experience explained by Calisi et al. is in line with the level of detail mentioned later on in Sect. 5, however, this paper does not go into how the Virtual Reality experience can be implemented and what method might be the best option to represent the environment.

2. **3D Visualisation of Egyptian Funeral Artefacts**
   In the paper presented by Caruso, Micoli, and Guidi (2015), they describe how they captured the 3D models of 4 different Egyptian artefacts through the use of a photogrammetric process and then implemented a Virtual Reality system using Unity (version 5), the Oculus Rift (DK2) and the Leap motion sensor in order to allow one to manipulate the virtual artefacts using their hands allowing for one to learn more about said artefacts. From their findings, Caruso et al. state that such a system was easy to use, albeit having its share of design flaws. They also state, however, that the system created could eventually be incorporated into a virtual implementation of the "path of the dead" where one would be able to re-enact the rituals that ancient Egyptians prepared their dead for their last journey [6].
   Although this paper predominantly talks about the visualisation of heritage artefacts rather than of heritage sites, it does indicate that the use of immersive Virtual Reality experiences within cultural heritage can help educate the general public through virtual interactivity and gamification. The author feels that such virtual interactivity and gamification of the learning experience would, in turn, lead to an overall improvement in the user experience when visiting cultural heritage sites.

4.1.3 The Local Scene—Laser Scanning and Visualisation of St Agatha's Catacombs

In his Doctoral thesis, Glenn Farrugia mentions the efforts done by Heritage Malta to produce 3D documentation of the St. Paul's and St. Agatha's Catacombs in Rabat, Malta [10]. The detail obtained within said documented 3D models use very similar techniques as those described by Calisi et al. and Barizi et al.

Farrugia mentions that the visualisation created for the Saint Agatha's Catacombs is more of a First Person Shooter implementation that allows one to traverse the 3D model, as was showcased in the 3D boot that was set up at the 2012 BMTA (*Borsa Mediterranea del Turismo Archaeologico*) at Paestum, Greece [10]. Although this representation is still an interesting one, it does raise the question if such a visualisation is user-friendly enough to allow everyone to use and experience the catacombs captured. It is the author's opinion that the implementation of a VR system to explore such catacombs is not only more immersive but also more user-friendly as well as a great tool to enhance one's user experience when visiting said sites.

4.2 Alternatives to Using 3D Models of Heritage Sites with Cultural Heritage

It is fair to say that the use of 3D models within Virtual Reality experiences does not necessarily have to be the only way in which VR can be used within cultural heritage. One of the main drawbacks of using such a system is that 3D modelling is time-consuming to produce and computationally expensive to render in real time. In their paper, Koeva, Luleva, and Plamen argue that more straightforward implementations such as the use of spherical panoramas can be just as effective, more user-friendly, more accessible as it can be implemented and viewed on a simple browser or smartphone and also more affordable [13].

One could, in turn, argue that such systems might not allow the user to learn more given that the views offered to the user are predetermined therefore it might not engage the user as much as a more interactive system.

4.3 Use of Virtual Reality with Sites that Are Very Difficult to Visit Physically

Virtual reality is a handy tool to use when considering very inaccessible locations. In their paper, Haydar et al. describe how VR in conjunction with Augmented Reality (AR) can be used to explore realistic reconstructions of heritage sites using immersion. In this case, they made of a real map representing a deep underwater site [11].

This is an important aspect of VR, as it allows one to be able to experience environments that one would not usually be able to access in typical situations.

5 Methodology of Creation of Artefact

As mentioned in Sect. 3, the overall motivations and goals for the creation of the proposed Virtual Reality Experience for the Saint Paul's Catacombs were:

1. To create an application that would make it easier and more intuitive to explore the 3D models of the catacombs.
2. To provide means by which a person with limited mobility would be able to explore said catacombs.
3. To make the experience as realistic as possible and
4. to have the user as immersed as possible.

The first step of the process was done by the company Digitarca s.n.c.. In July 2010, the company was commissioned by Heritage Malta to produce a 3D Laser Scanner survey of the entire St. Paul's Catacombs process. Once the survey was done, they

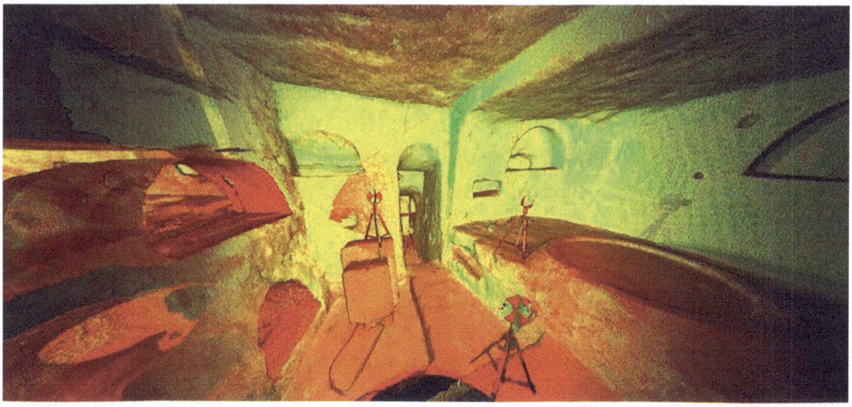

Fig. 2 A point cloud of one of the Saint Paul's Catacombs showing the level of detail obtained. *Source* http://www.digitarca.it/portfolio-items/catacombe-di-st-paul/?portfolioCats=65

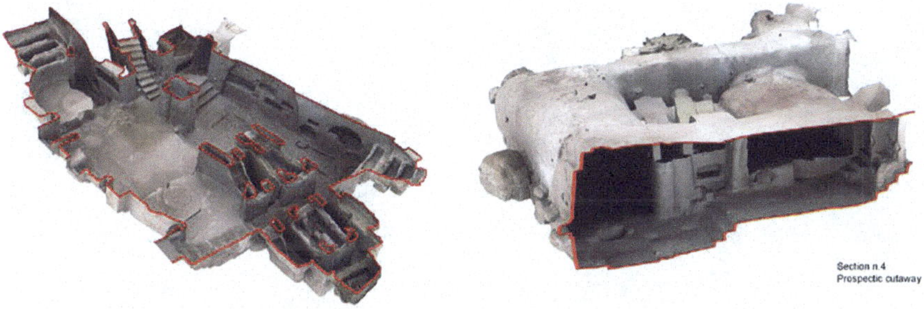

Fig. 3 3D models and sections of one of the St Paul's Catacombs. *Source* http://www.digitarca.it/portfolio-items/catacombe-di-st-paul/?portfolioCats=65

delivered several plans, sections and 3D models of the scanned catacombs [9]. An example of such scans can be viewed in Figs. 2 and 3.

These 3D models were then given to Saint Martin's Institute of Higher Education to use them to create the aforementioned Virtual Reality Experience. This experience was to be created using the Unity Engine (Unity 5) and the Oculus Rift Development Kit 2 (DK2) Virtual Reality headset. This was because, at the time (2014), the Oculus Rift DK2 was one of the only Virtual Reality headsets available, and Unity 5 offered full support with the Oculus SDK (Software Development Kit) [18].

5.1 PC Hardware Specifications Used

Since this VR experience needed to be as smooth as possible when running, it was essential to have a powerful enough computer to handle all of the computation and

Table 1 PC specs used for VR experience

Motherboard	ASUS X99-S
CPU	Intel Core i7 5820 K Six Core
RAM	Kingston predator 16 GB (4 × 4 GB) DDR4
Graphics card	MSI GeForce GTX 970 gaming edition 4096 MB DDR5 PCI-E
Hard disk	Seagate Barracuda 7200 RPM 1 TB SATA 6 Gb/s 64 MB Cache

Fig. 4 Screen shot of the FPS implementation. *Source* Author

rendering needed. Table 1 shows the specifications list of the PC built to run the experience.

5.2 A FPS VR Experience Implementation

The first implementation of the Virtual Reality Experience created, made use of a First Person Shooter (FPS) style controller which can be found implemented in a large number of computer games. Given the models generated by Digitarca, as well as plans of the locations and orientations of the catacombs by Heritage Malta, a set of scenes were created in which one can walk from one catacomb entrance to the next as well as have the opportunity to navigate the individual catacombs themselves. One can see a screenshot of the FPS VR experience in Fig. 4.

The way one explores the environment within this experience is through the use of an XBOX360 controller, as can be seen in Fig. 5. The user controls the camera

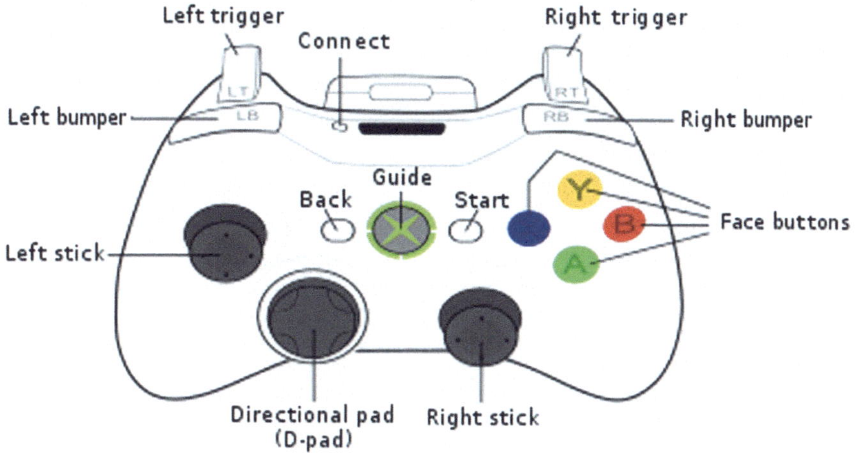

Fig. 5 XBOX360 controller with controls layout. *Source* https://en.wikipedia.org/wiki/Xbox_360_controller

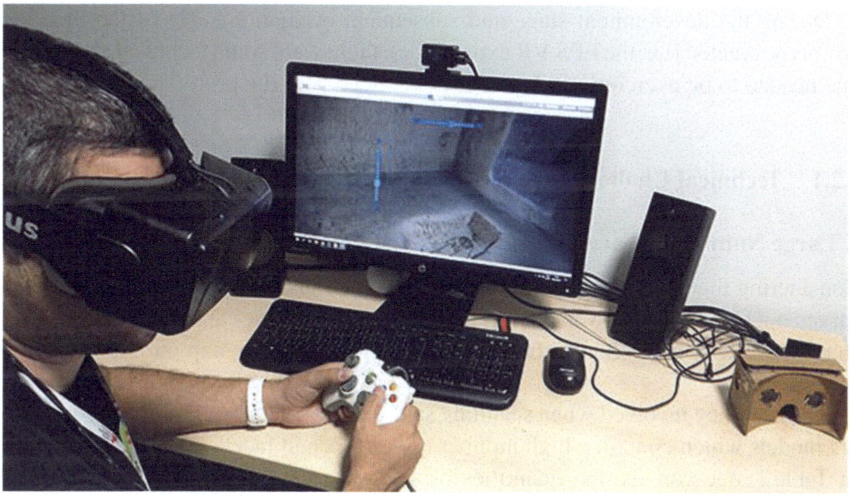

Fig. 6 The use of the Oculus rift with the second version of the VR experience. *Source* https://www.timesofmalta.com/articles/view/20170711/local/exploring-st-pauls-catacombs-in-virtual-reality.652986

through the use of either the direction pad or the left stick. Using these controls, the player can move forward and back as well as rotate left and right. Rotation of the view is also controlled through the Oculus Rift headset, which can detect head movement through an inbuilt gyroscope (not unlike those found in some mobile phones) with the aid of the depth sensor which can be seen on top of the monitor in Fig. 6. One also can "jump" in the experience by pressing the 'A' face button.

Table 2 Table showing the number of Vertices and Facets in each model used

3D Model for	Number of Vertices in the model	Number of Facets in the model
Catacomb 1	129,215	254,755
Catacomb 2	65,862	130,979
Catacomb 3	572,663	1,142,969
Catacomb 6	200,698	399,192
Catacomb 8	651,924	1,305,093
Catacomb 10	158,676	316,049
Catacomb 12	168,048	334,345
Catacomb 13	538,431	1,075,027
Catacomb 14	169,348	336,546
Catacomb 17	158,640	317,290
Catacomb 23	199,013	393,761
External	221,667	428,425

During the development stage and subsequent evaluation stage of the first VR prototype created (i.e. the FPS VR experience), there were some technical challenges that needed to be overcome and drawbacks noticed with the prototype developed.

5.2.1 Technical Challenges

A Large Number of Facets within the Models

Considering that the Saint Paul's Catacombs were constructed by having its spaces carved out from the rock by hand, it is understandable that the catacomb walls created were not exactly even; hence the need for laser scanning to map out the catacomb spaces.

The accuracy involved when scanning said spaces is very high; thus, it resulted in 3D models which contain a high number of vertices and facets (after optimisation). In Table 2, one can see the quantities of said vertices and facets for each model represented in the VR experience. This meant that since every frame needed to be rendered twice (one for each eye in the VR headset) a relatively high spec computer was needed to run the system smoothly. This is especially true given the fact that collision detection needed to be calculated on the model mesh itself.

Reversed Facets

One of the main issues encountered when developing this experience was that some facets of the 3D models were facing the wrong way. This is not that much of an issue when viewing the 3D model in a viewing software (such as Maya or 3D Studio Max) as when one renders a scene all facets[1] will be rendered. This is down to the

[1] At least those within the view space.

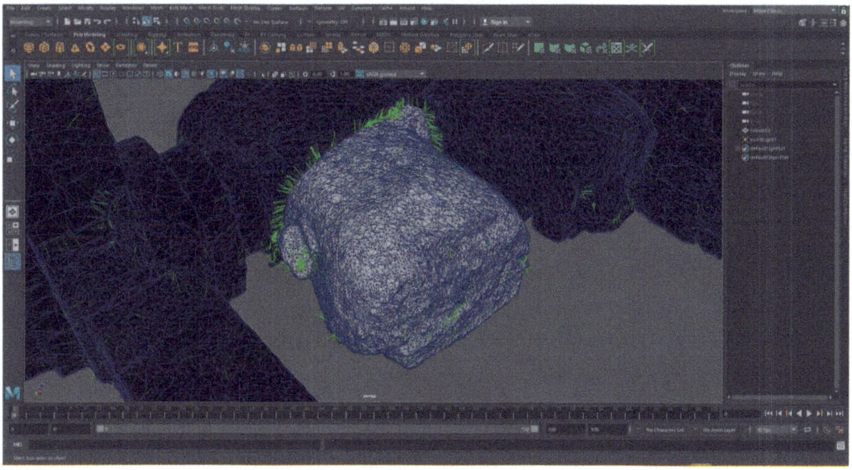

Fig. 7 Example of reversed faces in a Catacomb (14) model. *Source* Author

fact that such software does not need to render at Real-Time. An example of these reversed facets can be seen in Fig. 7. In this figure, one can see that there are two types of shaded surfaces; light and dark. The dark surfaces indicate that those facets are facing away from the perspective camera, while the light surfaces indicate that those facets are facing towards the camera.

Unity, by default, does not render facets which are not facing the camera. This is an optimisation to increase the rate at which a frame is rendered since Unity is typically used to develop video games. This meant that if one just placed the 3D models within a Unity scene, one would see several empty portions, which in turn would break the immersive power of the experience. With these reversed facets, it also meant that collisions would not work as expected because collisions, within Unity, would only be detected by a facet when an object is travelling towards said facet and when that facet is facing the oncoming object. To keep the frame rates as high as possible (at least 90 frames per second as indicated by the Oculus VR best practice guide [25]), the reversed facets were painstakingly altered back to their desired orientation using 3D modelling software.[2]

Connecting Multiple Meshes Seamlessly

Due to the size of the 3D files generated, it was necessary for one to split up interconnecting hypogea models into different files. However, it was necessary that these models would be reconnected in the VR experience. Given the fact that these models were large and also intricate, it was somewhat difficult to connect these models. This was because that one needed to first orient the models correctly and then make sure that they were correctly connected such that the joint is seamless. An example of this can be seen in Fig. 8.

[2]Primarily Maya and 3D Studio Max.

Fig. 8 Example of connected catacombs. In this case catacomb 1 and catacomb 2 (selected in orange). *Source* Author

Voids within the Meshes

In some cases, the 3D models given contained many voids or openings which if left within the VR experience, it might break the immersion of the user. There could have been several reasons as to why such voids existed, which is not the scope of this chapter to discuss. However, it was necessary to block off these voids so that immersion is kept intact. In Fig. 9, one can see an example of how these voids were blocked off. In this case, 3D cuboids were used covering up the voids within the model.

5.2.2 Drawbacks Noticed

Even though an FPS VR implementation to explore the catacombs is probably one of the more intuitive implementations, there were some drawbacks noticed with the design.

Fig. 9 Example of voids within the data provided. In this case the voids found between catacombs 1 and 2 (selected in orange). *Source* Author

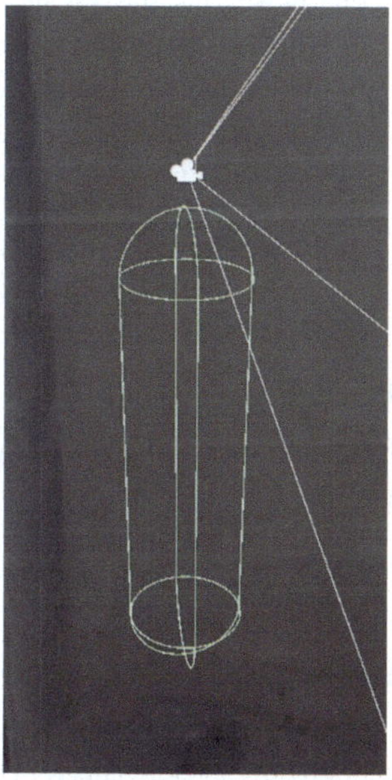

Fig. 10 The capsule volume of the player when exploring the Catacombs. *Source* Author

Movement Still Being Restrictive

The Catacombs included a lot of narrow passages, shallow ceilings, different stair heights, sudden drops, high and also small openings, etc. Due to these reasons, it is difficult for one to design a First Person Controller capable of navigating these spaces. This was because even though some spaces were made more available for one to explore when compared to the real-life catacombs, there were other areas which were still quite difficult to navigate within the VR experience.

One might argue that one could get around this by altering the player's capsule volume. Figure 10 showcases this volume. If one were to change the size and shape of this capsule volume, then it could be implemented in one of the two ways; either automatically or via input detected by the user. It is fair to say that both implementations have their shortcomings.

- **Automatically**
  If one had to implement automatic changes to the player's capsule volume as he/she arrives at specific points within the catacomb, then it might cause the camera to re-position itself. This re-positioning, if not expected by the user, might cause some disorientation, which in turn might lead to virtual sickness as explained below.
- **Via Input Detected by the User**
  If one had to implement the changes to the player's capsule volume via some input detected from the user, then it would mean that one would need to memorise the position and location of said commands on the input device (in this case that of the XBOX360 controller). This is because the user cannot see the controller when one is wearing the VR headset. This, in turn, poses several other problems. One of which is that if the user presses a wrong button thinking that they pressed the right button, then it might cause the system to cause some movement that the user was not expecting which again would lead to virtual sickness (again see below).
  The other problem that would arise is that if one is not an experienced game player, then it might cause the VR experience to be less user-friendly and therefore less useful as a tool in general. This option was therefore not taken as one of the design parameters of the project was to make the system as easy to use as possible to allow the maximum number of people to be able to use it.

Possibility of Getting Stuck in a Location

Keeping in mind the User Experience of the VR experience, it was decided that the player capsule would not change. Instead, it would be sized such that it would allow one to explore as much of the catacomb models as possible. This design decision did have a drawback in which it was possible to get stuck in a particular position within the Virtual Catacombs.

Possibility of Virtual Sickness

Due to the uneven layout of the Catacombs, it is quite possible for one to come across sudden dips and elevations in the level of the floor. This, in turn, would re-position the camera rather suddenly. The sudden movement of the camera can cause sensory (cue) conflict in which the brain. This conflict happens when one is visually receiving cues telling the user that he/she has moved while at the same time, the motion information received by the semicircular vestibular canals in the one's inner ears, is contradicting this information (i.e. telling the brain that no movement has occurred). This, in turn, can cause this so-called sensory conflict [20], which can lead to virtual sickness.

Virtual sickness is a pretty common phenomenon in the field of Virtual Reality, so much so that Oculus themselves provide several guidelines to reduce this unpleasant side effect. They state that ideally, the linear movement does not exceed a speed of 1.4 m/s, and they also recommend that one should not accelerate quickly and too frequently [25].

Inaccessible Locations

Although it is quite possible for one to explore much of the virtual catacombs, there were still some locations which were still hard, if not impossible, to reach. This was because the placements of these openings were in very awkward places.

5.3 *Drone VR Experience Implementation*

Due to the drawbacks noticed with the FPS implementation of the VR experience, it was decided that a second prototype would be created which would simulate the use of a Drone. Within this implementation, the user pilots a small virtual drone as can be seen in Fig. 6. The user controls are as follows:

- Forward movement through the forward control on the Direction Pad, a.k.a. D-Pad (see Fig. 5). Backward movement was omitted from this implementation.
- Left and right rotation of the drone is controlled through the use of the right thumb stick (Fig. 5)
- Rotation of the camera to view different perspectives is done through the movement of the head and hence the VR headset.

The interesting thing about this implementation is that one can "change" the altitude of the drone by just looking in the direction of where one wants to go and then press forward. Thus if one wants to climb, then one would need to look up, and if one wants to descend, then one needs to look down, all the while pressing the forward direction on the D-Pad. Other alterations made to this implementation include:

- **The Speed.** The speed of the character controller (i.e. the drone) was significantly reduced in an attempt to reduce the possibility of virtual sickness.
- **The inclusion of objects to find within the catacombs.** There were a few objects placed in and around the catacombs to make exploring the space more fun. The objects included were a Roman style oil lamp and a skeleton still laying in its original resting place.
- **A scan function.** The scan function allows one to emit a spherical pulse from the centre of the *drone* which keeps growing up to a certain distance. If the pulse intersects one of the objects placed within the virtual catacombs, then an information window would appear above the item describing what it is.

6 Evaluation

In this Section, we shall be comparing and contrasting the User Experience of someone exploring the Saint Paul's Catacombs with and without the added introduction of the developed VR experiences. First, we shall take into consideration a typical visit without the VR exploration.

6.1 Evaluation of the User Experience of a Visit to St. Paul's Catacombs Excluding VR Exploration

When evaluating the user experience of any process, one needs to take into considerations all of the different steps one needs to or can do within said system. In general, there are several milestones that one needs to accomplish to have a successful user experience. It is fair to say that there can be some factors (internal or external) that can affect the outcome of the user experience.

Learning of the Site's Existence

For a user experience of the Saint Paul's Catacombs to start, first one needs to hear of its existence. If one does not know about it, then it is doubtful that he/she would become a visitor to the site, unless one stumbles across it accidentally and his/her curiosity is peaked enough to visit the said site. One can learn about the site in various ways, such as:

- Websites—such as Wikipedia, Trip Advisor, Blogs, Travel Websites, Google, etc.,
- Adverts—TV adverts, Web adverts, Facebook Ads, etc.,
- Literature—Travel Guides, Magazines, Newspapers, etc.,
- Word of mouth.

It could be that the user might find it difficult to learn about the site and what one should expect when visiting it. It could be that the references to the site are very obscure, so much so that it is difficult to find out any information about the site at all.

Finding the Catacombs

After one learns about the site, the next step is for one to find where the site is and how to get there. It follows that for one to start their exploration of the Saint Paul's Catacombs, one would first need to find where these catacombs are. Again, at this stage, one can experience a positive or negative outcome.

A positive outcome is experienced if the site is easy to find. One might opt to use:

- freely available maps which give clear instructions as to how to get there
- Google maps or a GPS
- clear road signs, etc.,

On the other hand, a negative outcome is experienced if one cannot find the site. For example, the marker on Google Maps is not accurate and therefore useless to whoever wants to use it. Another example can be confusing road signs, which can lead to the user getting lost.

Buying an Entrance Ticket

Once one reaches the site, the next step is for him/her to purchase a ticket to access the complex. In this situation, a positive outcome would be one in which one can purchase the ticket using a payment method that they have access to while being greeted in a friendly and professional manner from the ticket station attendant.

Conversely, a negative outcome might come up through lack of payment options; for example, having the site accept only cash and not credit cards. Another example might be if the complex does not make use of digital wallets such as Google Pay or Apple Pay.

Exploring the Complex

Once a ticket has been purchased, one can start exploring the complex. In this case, a positive experience is one in which one would be able to easily navigate from one exhibit to the next with the use of helpful and clear signposts.

The Saint Paul's Catacombs complex only allows one to navigate from Catacomb to Catacomb outside (i.e. not sheltered). Therefore one would always be exposed to the elements. This too can affect the user experience.

Exploring the Catacomb

When exploring the Saint Paul's Catacombs complex, one would eventually come across entrances to the different catacombs themselves. Here, the user experience depends on several factors:

- Are the catacombs well lit, or are they too dark to see?
- Is there any interesting trivia that one can learn when in the catacombs?
- Can one physically access parts (if not all) of the catacombs?
- Are there any limitations as to where one can explore, or is one allowed to roam freely?

Notice that once one has explored a catacomb, then he/she would be able to explore the complex and any other catacombs as many times as they so wish.

Leaving the Complex

At some point, one is going to wish to end their experience within the Saint Paul's Catacombs complex and hence would start looking for the exit. In this case, the user experience depends on how easy it is to find the exit or possibly the gift shop. The easier it is to find, and the closer one already is to the exit once they have reached the endpoint of their experience, the more positive the experience at that stage is going to be.

It might be the case that one would want to share their experience on social media or review sites. In turn, this might help with the other people getting to know of the

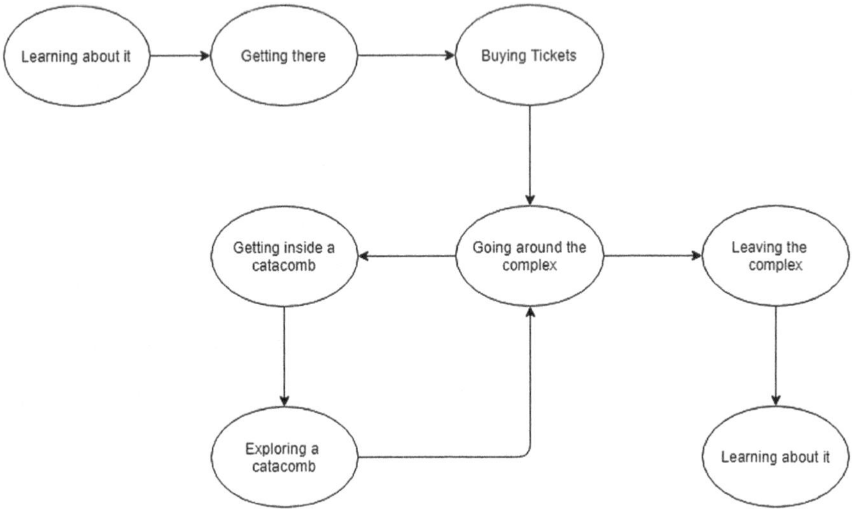

Fig. 11 Current flow of experience. *Source* Dylan Seychell

Saint Paul's Catacombs. This experience would be further enhanced if one is allowed to take photographs within the complex itself, thus encouraging one to do so (at least eventually).

Flow of the Experience

If one had to map the user experience flow when visiting the Saint Paul's Catacombs as described above (i.e. without introducing the Virtual Reality Experience created), one would end up with something like the diagram shown in Fig. 11.

6.2 Evaluation of the User Experience of a Visit to St. Paul's Catacombs Including VR Exploration

Lets re-evaluated the user experience of one's visit to the Saint Paul's Catacombs. However this time we shall also include the Virtual Reality Experience described in Sect. 5.

As discussed in Sect. 3, the Virtual Reality experience's goals were not just about accessibility (or people with mobility issues). The project was also designed to explore difficult to reach areas, allow more people to experience more of the catacombs an also to experience the catacombs in their most essential form (i.e. without being obstructed by modern pathways or lights and cables.

Removing Barriers

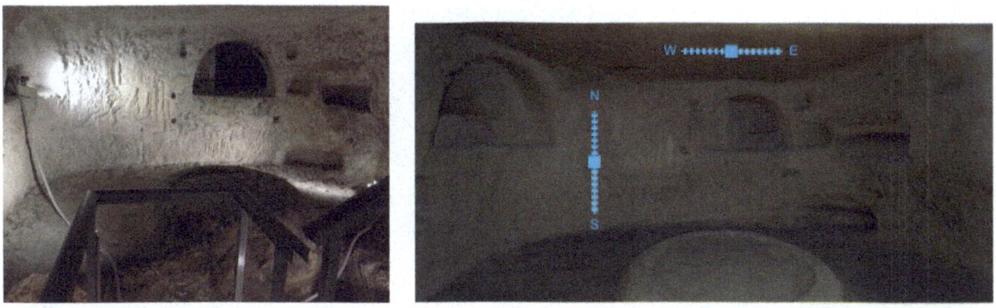

(a) View inside catacombs with physical barrier

(b) View inside Drone VR implementation

Fig. 12 A comparison of boundaries in the actual catacombs versus with the Drone VR implementation. *Source* Dylan Seychell

In Fig. 12, one can see a photograph of one of the catacombs at the Saint Paul's Catacombs complex and a similar point of view of the virtual version of the same Catacomb. It is evident that within Fig. 12a there is a physical barrier which is there to stop visitors from venturing any further from the handrail, thus limiting the spaces that can be explored.

This same limitation, however, is not present within the Drone VR implementation, as seen in Fig. 12b. This shows that the Drone VR implementation can allow the user to explore more of the catacombs than one might be allowed to under normal circumstances. This is but an example as to how the VR experience can complement the real world exploration of the catacombs. The VR experience allows one to explore more of the catacombs without having to worry about barriers and physical obstacles.

Giving Access to Hard to Reach Areas

Some chambers within the catacombs are very hard to reach as their openings can be in several awkward locations. Also, some of these said openings are relatively small and therefore might be unsafe for one to explore as it might cause someone to get stuck. One such opening can be seen in Fig. 13a. This opening can be found on top of a doorway, which is itself leading down to a deeper chamber. The opening is relatively small in diameter. Given all of these facts, one can say that the openings are largely inaccessible, even to people who have no mobility issues.

Using the Drone VR implementation, one can very easily access this space virtually. Since one is controlling a small drone, all one has to do is look in the direction of the opening and 'fly' towards it. This can be seen in Fig. 13b. Again, this is but an example as to how the Drone VR implementation can provide access to all users.

Allowing One to View More Detail

Some walkways and stairways at the Saint Paul's Catacombs have been covered over by metal paths in an attempt to preserve these pathways from erosion due to heavy use from visitors to the site. Although this measure does protect the pathways as

(a) An opening just above a doorway

(b) The same opening just above the doorway in Drone VR implementation

Fig. 13 A comparison of accessibility between real world catacomb and the Drone VR implementation *Source* Dylan Seychell

mentioned above and stairways, they do cover up the original construction that has been there since the 7th or 8th centuries AD [14]. An example of a covered stairway can be seen in Fig. 14a.

By contrast, Fig. 14b showcases the same staircase; however, minus the modern day protection. This allows a user of the VR experience to view the catacombs in their original (unmodified) state. Being able to do so allows one to be able to view more detail which might have been lost had it not been for the virtual catacombs experience to complement the real world one.

6.3 Improved Flow of Experience

From the comparisons made above, it is clear that the VR experience created, complements the overall user experience that a visitor can obtain when visiting the Saint Paul's Catacombs. As mentioned earlier, the VR experience allows one to be able to explore and see things about the catacombs, which would be near impossible (or at least very hard to do) without it. Making use of Virtual Reality experiences to complement real-world experiences can help give users the ability to go beyond just scratching the surface, especially when implementing it to improve awareness of the cultural treasures offered in Cultural Heritage sites. Such systems can allow users to learn more of the through their exploration, which in turn fortifies one's learning done when viewing the real world artefact. Figure 15 showcases how this VR experience integrates itself with the real world experience shown in Fig. 11.

(a) Metal staircase covering original staircase

(b) VR implementation without metal staircase

Fig. 14 A comparison of availability to view original details between the actual catacombs versus the Drone VR implementation. *Source* Dylan Seychell

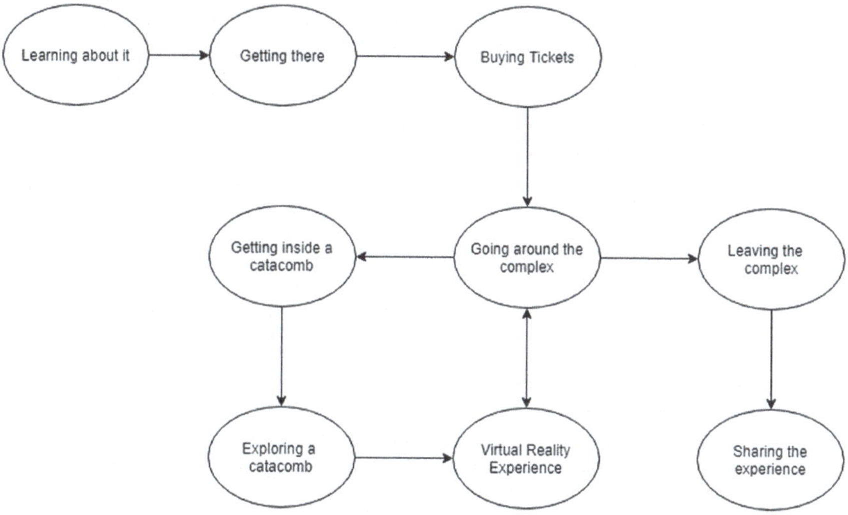

Fig. 15 Improved flow of experience. *Source* Dylan Seychell

6.4 *Preliminary User Feedback of Experience*

On the 24th and 25th of March 2017, there was a Technology and Gadgets Expo coined the Malta Robotics Olympiad (MRO) which was held at the MFCC in Ta' Qali, Malta. The MRO is the largest technology expo that occurs on the Maltese

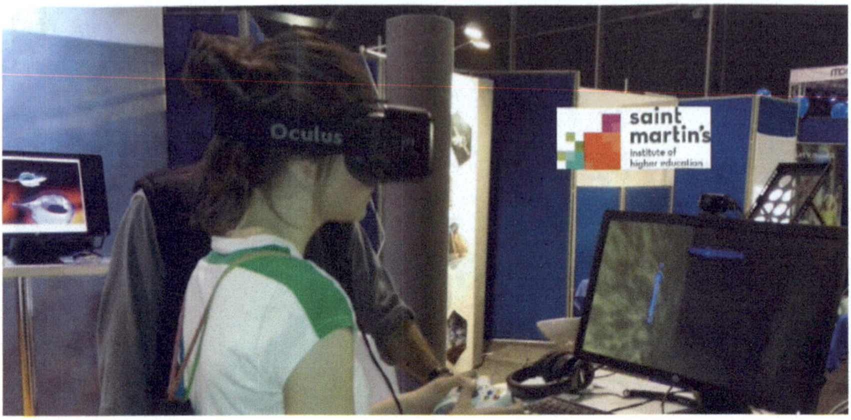

Fig. 16 Drone Experience field tested at MRO. *Source* https://lovinmalta.com/lifestyle/technology/you-can-soon-go-on-a-virtual-reality-tour-of-maltas-catacombs

islands. During this event, the general public was given its first opportunity to view firsthand the developed drone VR experience of the Saint Paul's Catacombs.

On the whole, the experience created was positively received, with several people commenting positively on the immersive feeling that they experienced as well as the quality of the visuals presented. As with most artefacts of this nature, there were some criticisms voiced (Fig. 16).

People who do not have much experience with the XBOX360 controller (or similar types of controllers), found it difficult to grasp how one should control the virtual drone, however with some guidance this issue was addressed. One could speculate that since one is not able to glance at the controller while using the experience, it is easier for one to become disorientated or befuddled with the controller. Later versions of the Oculus Rift addressed this issue as one was given the ability to see a virtual pair of hands showing which buttons are being pressed/manipulated on the new so-called touch controllers provided (Fig. 17).

Some people posed questions relating to why no backward movement was implemented for this system. This feature suggestion was requested repeatedly, so much so that steps were taken to modify the drone controller created in order to incorporate it.

Others still asked why separate parts of the XBOX controller controlled the forward movement and the rotational movement of the virtual drone. This was primarily asked by people who had quite a lot of videogame experience and therefore were well versed in popular movement controls provided in other gaming experiences. This question was discussed between the developers (i.e. Mark Bugeja and myself). It was concluded that having all movement controls on, say the D-Pad or the left analogue stick (as is used in most games), would result in a lack of precision when controlling the virtual drone.

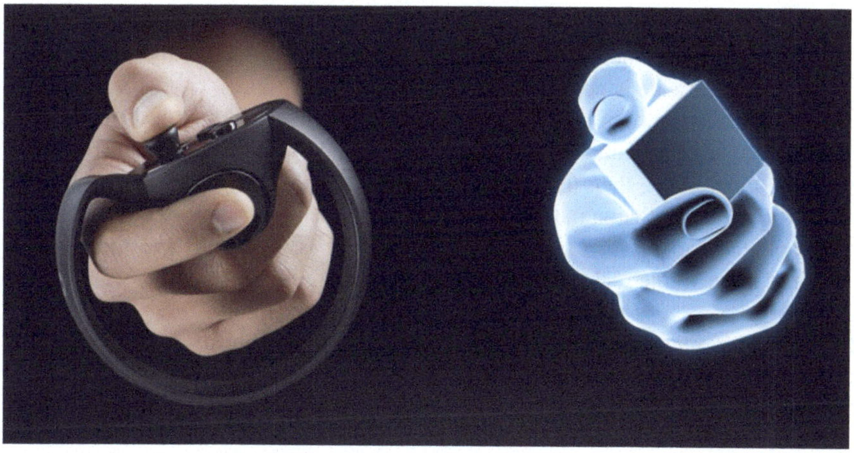

Fig. 17 Oculus Touch controller and virtual representation. *Source* https://www.electronicsweekly.com/blogs/gadget-master/consumer-electronics/tearing-oculus-rift-touch-controller-2017-01/

This lack of precision would arise from the nature of the D-pad and analogue stick—these control devices allow for one to enter more than one command at a time. Example: move forward and rotate at the same time. This property thus makes these input devices less precise compared to their expectations of what the drone should be doing. If the user thinks that he/she requested the drone to move in a particular way (example moving forward) and the drone rotated as well as move forward it would cause user frustration and confusion. The frustration causes would hence decrease the spatial presence [23] within the experience and possibly reducing how immersed the user would feel. This inconsistency or confusion caused between expected movement vs actual movement could also result in virtual sickness as these unexpected movements can result in the player feeling disorientated [22].

7 Conclusion and Future Work

In Sect. 3, the following objectives were set for the Virtual Reality Experience of the Saint Paul's Catacombs. One can also say that the VR experience has managed to reach these goals for the reasons mentioned below:

1. **To create an application that would make it easier and more intuitive to explore the 3D models of the catacombs**
   The nature of camera control in most 3D model viewing software, one would be expected to navigate the 3D model by manipulating the camera with 3 types of controls being panning, rotating (about a point somewhere in front of the camera) and zooming. This is not particularly a user-friendly way of how one can navigate a scene. This kind of camera movement is not that intuitive to use, especially for

people who have never used such software. Also, this software does not take into consideration that the camera cannot pass through the model. This is not surprising as the software is not designed to do so.
With this respect, one can say that the VR experience created has managed to achieve this goal. Camera rotations are controlled by head movement (Example: if one looks right, then the camera would turn right, etc.). The VR experience also considers collision meaning that one cannot simply pass through walls. This means that the way that one navigates the scene is very similar to the way one would move in the real world, thus making it more intuitive.

2. **To provide means by which a person with limited mobility would be able to explore said catacombs**
The fact that the VR experience allows one to sit down and explore the catacombs through the use of a head mounted display and a game controller means that, for the first time, people who physically cannot get to the catacombs to see them for themselves, can now have a realistic glimpse of what these spaces are like. Moreover, this system provides improved access to all visitors as it allows one to explore the catacombs without barriers, with the ability to reach places which are inaccessible in real life and with more original detail than one can see by physically being there. One could also argue that this also answers the research question that was posted in the beginning being:

 > In the case of the Saint Paul's Catacombs, can an implementation of the catacombs environment in virtual reality add value to the user experience when one is exploring said heritage site? Moreover, if so, how can this value be added?

3. **To make the experience as realistic as possible and to have the user as immersed as possible**
Considering that the 3D model used was created using 3D laser scanners and that the textures where captured using high-quality photography, and that the system is running on a relatively powerful computer[3] (at least for the time it was created, i.e. 2015) one can say that the visuals created are quite realistic. Since the VR experience was designed to allow one to explore the catacombs at one's own pace taking the paths that the user decides to take, one can argue that it is more immersive than a pre-rendered tour in which the user is just an idle spectator.

This research, like all other pieces of research that came before it, is by no means complete. It would be nice if the research presented here is expanded upon by creating a qualitative analysis of the artefact to verify the effectiveness and quantify the possible improvement made of the user experience on site (i.e. at the Saint Paul's Catacombs complex). One might also try to see if a similar implementation would be useful in other heritage sites such as the Hypogeum in Hal Saflieni, Malta (the very fragile nature of that Neolithic site where the humidity, light, number of visitors and temperature values are all strictly controlled so as not to damage it and the red ocker paintings and patterns that can be found on its ceilings).

[3] See Table 1.

Another field of study in which one could use a similar system is that of experimental virtual archaeology. Experimental virtual archaeology in which one can learn new facts about the site by conducting experiments within the virtual environment, which might be challenging to conduct otherwise. Another opportunity for study might be the use of similar systems for educational purposes; possibly through the implementation of a gamified version of the VR experience. Another possible addition to a similar implementation is that of adding some form of narrative showcasing historical events, or eve re-enactments of that could have happened when such sites were used, and then studying how useful such an implementation is in terms of engagement and learning.

References

1. M. Bekele, R. Pierdicca, E. Frontoni, E. Malinverni, J. Gain, A survey of augmented, virtual, and mixed reality for cultural heritage. J. Comput. Cult. Herit. (JOCCH) **11**(2), 1–36 (2018)
2. F. Biocca, M.R. Levy, Communication in the age of virtual reality, *Virtual Reality as a Communication System* (Erlbaum Associates Inc, Hillsdale, 1995), pp. 15–31
3. M. Brizzi, S. Court, A. d'Andrea, A. Lastra, D. Sepio, 3d laser scanning as a tool for conservation: the experiences of the herculaneum conservation project (2018)
4. D. Calisi, F. Cottefoglie, L. D'Agostini, F. Giannone, F. Nenci, P. Salonia, M. Zaratti, V.A. Ziparo, Robotics and virtual reality for cultural heritage digitization and fruition. Int. Arch. Photogramm. **XLII-5/W1**, 503–508 (2017)
5. F. Cameron, S. Kenderdine, *Theorizing Digital Cultural Heritage: A Critical Discourse (Media in Transition)* (The MIT Press, Cambridge, 2007)
6. G. Caruso, L. Micoli, G. Guidi, 3d visualization of cultural heritage artefacts with virtual reality devices, vol. xl (Copernicus GmbH, Gottingen, 2015), pp. 165–172
7. D. Cassar, St paul's catacombs restoration and new visitor's centre inaugurated - tvm news (2017), https://www.tvm.com.mt/en/news/st-pauls-catacombs-restoration-and-new-visitors-centre-inaugurated/. Accessed on 12 Oct 2018
8. M. Christofi, C. Kyrlitsias, D. Michael-Grigoriou, Z. Anastasiadou, M. Michaelidou, I. Papamichael, K. Pieri, A tour in the archaeological site of choirokoitia using virtual reality: a learning performance and interest generation assessment, in *Advances in Digital Cultural Heritage*, ed. by M. Ioannides, J. Martins, R. Žarnić, V. Lim (Springer International Publishing, Cham, 2018), pp. 208–217
9. Digitarca, Catacombs of st. paul - type (2018), http://www.digitarca.it/portfolio-items/catacombe-di-st-paul/?portfolioCats=65. Accessed 10 Dec 2018
10. G. Farrugia, The presentation and interpretation of early Christian heritage in Malta: Past, present and future (2017)
11. M. Haydar, D. Roussel, M. Maïdi, S. Otmane, M. Mallem, Virtual and augmented reality for cultural computing and heritage: a case study of virtual exploration of underwater archaeological sites (preprint). Virtual Real. **15**(4), 311–327 (2011)
12. C. Holtorf, G. Fairclough, *The New Heritage and Re-Shapings of the Past*, 1st edn., Archaeological Orientations (Routledge, 2013), pp. 197–210
13. M. Koeva, M. Luleva, P. Maldjanski, Integrating spherical panoramas and maps for visualization of cultural heritage objects using virtual reality technology. Sensors **17**(4) (2017)
14. H. Malta, St. Paul's catacombs (2018), http://heritagemalta.org/museums-sites/st-pauls-catacombs/. Accessed 07 Dec 2018
15. H. Malta, Who we are (2018), http://heritagemalta.org/about-us/who-we-are/. Accessed 7 Dec 2018

16. H. Malta, Annual report 2009 (2009), http://heritagemalta.org/wp-content/uploads/2012/03/Annual-Report-2009-Final.pdf. Accessed 7 Dec 2018
17. H. Malta, Heritage malta and st. martin's institute of it sign memorandum of understanding (2013), http://heritagemalta.org/heritage-malta-and-st-martins-institute-of-it-sign-memorandum-of-understanding/. Accessed 8 Dec 2018
18. Oculus, Oculus utilities for unity (2018), https://developer.oculus.com/downloads/package/oculus-utilities-for-unity-5/. Accessed 10 Dec 2018
19. D. Pletinckx, Virtual archaeology as an integrated preservation method. Virtual Archaeol. Rev. **2**(4), 33–37 (2011)
20. M. Potel, Motion sick in cyberspace. IEEE Comput. Graph. Appl. **18**(1), 16–21 (1998). Jan
21. H. Rua, P. Alvito, Living the past: 3d models, virtual reality and game engines as tools for supporting archaeology and the reconstruction of cultural heritage – the case-study of the roman villa of casal de freiria. J. Archaeol. Sci. **38**(12), 3296–3308 (2011)
22. J. Seibert, D.M. Shafer, Control mapping in virtual reality: effects on spatial presence and controller naturalness. Virtual Real. **22**(1), 79–88 (2018). Mar
23. P. Skalski, R. Tamborini, A. Shelton, M. Buncher, P. Lindmark, Mapping the road to fun: natural video game controllers, presence, and game enjoyment. New Media Soc. **13**(2), 224–242 (2011)
24. UNESCO (2018) Maltese catacomb complexes - unesco world heritage centre (2018), https://whc.unesco.org/en/tentativelists/1113/. Accessed 8 Dec 2018
25. R. Yao, T. Heath, A. Davies, N. Mitchell, P. Hoberman, Oculus vr best practices guide (2014)

Techniques of Filming and Audio Recording in 360-Degree Ambiences

Joseph Camilleri

Abstract Education has always been an important factor of life, being continuously analysed, in the attempt to improve its delivery in today's classrooms. Although much has been done to give education more interesting ways of delivery, yet there are several generic instances when educational techniques used in today's classrooms are deemed as outdated, both by educators and their students. Subjects, which are meant to enhance the knowledge and appreciation of a culture's heritage, can at times not be exposed to students in the most exciting way possible, so as to enhance learning and maximise understanding. A country's heritage is the map to its history. The accumulation of its languages, including its artistic endeavours and representations, stand as a reminder of our ancestors who have toiled hard to create the story that we are nowadays striving to keep alive and further enrich through contemporary means. Technology has become a tool which stands alongside the brushes and rasps of the artists and sculptors of antiquity. Today, computers and their burgeoning peripherals have given art newer twists and further methods of expression, which can in turn augment the way students are drawn into the magical world of their country's heritage. This project is endeavouring to capture film language and transpose it into a 360-degree film environment, which combined with the enrapturing use of spatial sound will recreate an epic moment in the fairly unknown initial stages of the Great Siege of Malta. This immersion is aimed not only to excite the young minds of students through the narrative techniques used, but further create compassion through an increased sense of empathy.

1 Introduction

Film has been one of the greatest invention of the 19th, 20th and the 21st centuries. The fact that film has existed for so long is proof enough that there was, and still is, something magical in films, which humans cannot seem to find elsewhere. Or if

J. Camilleri (✉)
Saint Martins Institute of Higher Education, Hamrun, Malta
e-mail: jcamilleri@stmartins.edu

D. Seychell and A. Dingli (eds.), *Rediscovering Heritage Through Technology*, Studies in Computational Intelligence 859,
https://doi.org/10.1007/978-3-030-36107-5_9

it does exist elsewhere, then such an element still cannot replace film entirely or in part.

The fact that film has remained attractive for a large part of society definitely brings forward the idea that there is something special about it. This translates in the fact that film is alive. It was born in the late 1800s with experimentations by the likes of the Lumiere brothers, Georges Melies and Alice Guy-Blache [1], all considered as the pioneers of film language. Film developed through the works and experiences of a myriad of directors like Alfred Hitchcock, Martin Scorsese and Akira Kurosawa, just to mention a few.

The end of the 20th century brought about individuals like Jaron Lanier, a VR pioneer, aptly referred to as the "father of virtual reality", who, together with other individuals declared the advent of a strong development within traditional film. This was the birth of Virtual Reality (VR) and the 360-degree camera. Although the traditional camera has become obsolete for these new ways of narrating stories, yet the film language, developed over a century of practice, has bridged its way over from one medium to the other.

Human beings are always searching for a better way to be able to express their individuality and the coming of the digital age, following other developments throughout history, such as the industrial revolutions, the mechanical age and the age of the Internet, has signified a whole new set of developments which have helped in the unlocking of the myriad of data contained in our structure. Klinke and Surkemper [2], describe this revolution as having a vast effect on society as a whole.

> Financial transactions happen in real time, global communication via the Internet is available free of charge, and the smartphone is our ubiquitous companion [2].

The digitising of a great amount of information, hence making it available for discourse on a worldwide scale, has increased opportunity to reflect on such occurrences and the possibility to experiment with the meanings that are extracted.

> We want to take part in developing and strengthening collaborative work internationally and interdisciplinary and — amongst other things — bringing (art) historians and computer scientists together [2].

2 Background

Bishop [3] describes the notion of derive. This is described as a tool in "psychogeography", which studies the effects of a given environment on the emotions and behaviour of individuals. The notion of derive acts as a mode of increasing one's awareness of urban surroundings. In the case of Situationist "unitary urbanism", derive enabled an undoing of the dehumanising effect of modernist forms of contemporary urban high-rise living [3].

360-degree filming projects reuse these concepts in trying to understand the somewhat insensitive effects of collective education, through the implementation of an experience which will attempt to isolate instances in a human's existence and portray the reality that extends beyond the visual through the provoking of empathetic situations in the individual's life.

Bishop also notes the conflict between art and the audience in the realm of participatory art. She emphasises the possibility of understanding art without the audience as seen in past artistic development—Bishop heralds an art wherein everyone is capable of producing.

> Indeed, in its strictest sense, participation forecloses the traditional idea of spectatorship and suggests a new understanding of art without audiences, one in which everyone is a producer. At the same time, the existence of an audience is ineliminable, since it is impossible for everyone in the world to participate in every project [3].

Martin Heidegger [4], in his essay "The Question Concerning Technology" made assertions which renovated the relationship of the artist with his/her materials and tools [5]. An example which can illustrate this point is "Bourne Identity"s' director Doug Liman's five-episode miniseries called "Invisible", which was adapted from 2D film scripts hence producing a result which to many seemed disorienting [6]. Heidegger asserted that relationships between artists and their tools need to be revisited to enable a further and more in-depth understanding of the possibilities of the tools at hand and the way forward for the artist [5]. Liman comments,

> The thing that became most clear to me when we first shot the tests was that we had to rethink the way we were telling stories, because when you just take a traditional scripted scene out of any TV script or movie script and shoot it in 360-degrees, it's going to be less compelling than what was shot in 2D [5].

Unlike games, where players can take their time to play, 360-degree filmmakers need to tell their stories in minutes. Compared to traditional filmmaking, 360-degree filming needs to be rethought on a new level. This would mean the application of existing techniques from traditional filmmaking to new techniques, born through the adaptation of this new means of narrative technology. Given that the viewer has the ability to look around, a director cannot simply change shots, this would create the possibility of getting the viewer to lose the sense of where he or she is. Filmmakers cannot use conventional strategies and methods to film a scene, as the scene is a whole bubble with no beginning and no end, says Simon Robinson from the software firm, The Foundry. The immersion created once a headset is donned would be broken once any flaws in filming are sensed [7].

Figures 1 and 2 demonstrate visually, how technology had an important role in the increase of immersion, or as Maurice Merleau-Ponty calls it—the Chiasm [9]. According to Mark Billinghurst [8] in his keynote speech given at the VSMM,[1] realism has increased with the passage of time and the invention of new technologies which allowed the viewer to partake more in the experience. With the arrival of 360-degree video, and the involvement of an actual constructed 3D space in the 2000s, reality kept on soaring upwards as it attained a more virtual application.

[1]The 22nd International Conference on Virtual Systems and Multimedia, http://www.vsmm2016.org/.

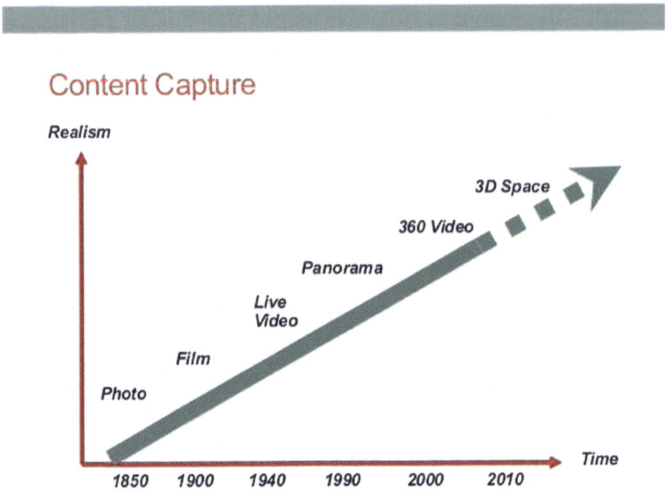

Fig. 1 Keynote talk given by Mark Billinghurst at the VSMM 2016 conference on October [8]

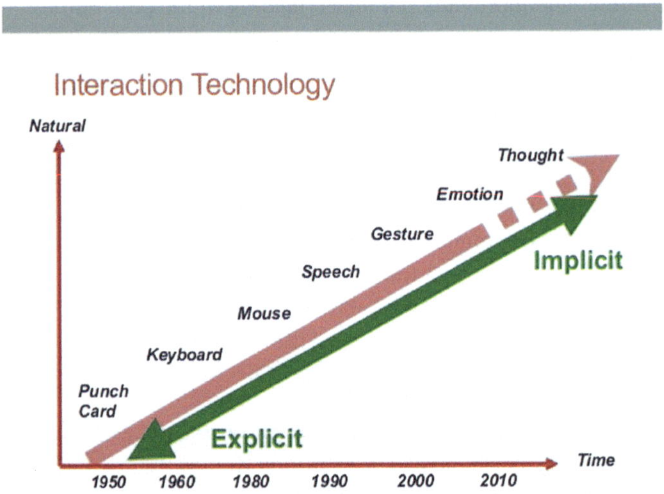

Fig. 2 Keynote talk given by Mark Billinghurst at the VSMM 2016 conference on October [8]

With time, more physical and psychological reactions have been employed through the engagement of new technologies which invited the use of such physical and psychological responses, such as gesture, emotion and thought. Would Billinghurst then be saying that prior to the 2000s, the viewer was not employing gesture, emotion and thought whilst experiencing paintings and/or film? The truth is that it is difficult to word the difference between; immersion in, for example a High Renaissance or Baroque painting and immersion in a painting done wholly using technological and virtual means, like the Tilt Brush for HTC Vive and recently also adapted for Oculus

Rift [7], and immersion in the direction of a film projected entirely in 360-degrees. All three have factors of immersion which are similar and will continue being similar, as these are the factors which are intrinsic in all art of a particular aptitude. Other factors of immersion would be singular to the medium being used for that specific artwork. To better understand these last statements one can quote Martin Heidegger in his essay, "The Question Concerning Technology" [4], wherein he states,

> Technology is not equivalent to the essence of technology. When we are seeking the essence of "tree", we have to become aware that what pervades every tree, as tree, is not itself a tree that can be encountered among all the other trees [4].

Other attempts in early filmmaking played upon the concept of "shock" and its effect upon the viewers, which were seated in a position of observers. Yet, this new medium of portraying images still served to transgress what was normality at the time. The Lumière brothers, French inventors and pioneer manufacturers of photographic equipment, devised an early motion-picture camera and projector called the Cinematographe, the precursor of today's cinema [10]. Hellmuth Karasek describes this episode of history very eloquently, when he says,

> One short film had a particularly lasting impact; yes, it caused fear, terror, even panic ... It was the film "L'Arrivee d'un train en gare de la Ciotat" (Arrival of the Train at La Ciotat Station) ... Although the cinematographic train was dashing toward the crowded audience in flickering black and white (not in natural colors and natural dimensions), and although the only sound accompanying it was the monotonous clatter of the projector's sprockets engaging into the film's perforation, the spectators felt physically threatened and panicked [10].

Having never experienced something similar in effect brought the viewers to a collective emotion of panic initiated through what they were seeing and actually "believing".

In his endeavour to build the "Empathy Machine", Milk asks the question,

> But then I started thinking, okay, well that's a part of you, but how do I put all of you inside of the frame? [11]

and embarked on the production of art installations which transposed the being of the user/viewer into the experience itself [12]. His attempts to create ultimate empathy brought him towards trying to transpose the user/viewer further into the created world itself. That is not seen in the window of the new reality, as in most applications using Augmented Reality, but

> through the window ...on the other side, in the world, inhabiting the world [11].

Milk describes 360-degree film as an "experiential medium", being inside the machine is another truth, as one would feel

> present in the world that you're inside and you feel present with the people that you're inside with ...[11].

In a quest to answer his final question brought him to opt for completely different filming equipment from the traditional.

Fig. 3 Screenshot of 1:2, 360 degree footage of the production, "Clouds over Sidra" [11] (United Nations Virtual Reality, n.d.)

> So now I've got you inside of the frame, and I saw people having even more visceral emotional reactions to this work than the previous one. But then I started thinking about frames, and what do they represent? And a frame is just a window. I mean, all the media that we watch—television, cinema—they're these windows into these other worlds. And I thought, well, great. I got you in a frame. But I don't want you in the frame, I don't want you in the window, I want you through the window, I want you on the other side, in the world, inhabiting the world …So that leads me back to virtual reality …It's a machine, but inside of it, it feels like real life, it feels like truth …So, it's a machine, but through this machine we become more compassionate, we become more empathetic, and we become more connected. And ultimately, we become more human [11].

In a joint production with Gabo Arora, Milk enters the world of a twelve-year-old in the Za'atari camp in Jordan—home to 84,000 Syrian refugees. It follows the girl to school, to her makeshift tent and even to the football pitch [11]. Producers Here Be Dragons presented "Clouds Over Sidra" (see Fig. 3), a 360-degree film for the United Nations in partnership with Samsung. The film was commissioned as part of the UN's advocacy at the World Economic Forum in Davos to state and business leaders and offers a unique perspective into the lives of the world's most vulnerable people [11].

> Getting a sense of "presence" from virtual reality and 360-degree filming is an incredibly powerful experience. Doing so in such a heart-breaking place will, we hope, have a profound impact on the viewer that traditional newsreels and film making has so far failed to convey. There is an urgency and visceral immersion to the film" said Patrick Milling Smith of "Here Be Dragons" [11].

360-degree filming is described as a means to connect humans to other humans in a profound way, previously unseen in any other form of media, hence changing peoples' perception of each other, creating a potential to change the world. Thus, this relatively new technological development—360 film, is a machine which connects individuals through empathy [11]; by placing the viewer inside a situation developed by the film director and not simply as an onlooker, observing actions but as a participant who can have an effect on proceedings in the experience, or be affected by the proceedings in the experience.

Patrick Milling Smith, producer of "Clouds over Sidra" [11], describes traditional film as wanting in enough potential and as inspiration to generate empathy in an audience. He labels VR and 360 film as the ultimate tool which goes beyond traditional methods. Milling Smith sees VR and 360 film as a new creation, unlike Milk whose work determined VR and 360 film as a technology which uses narrative in a diverse way from traditional filmmaking, yet could not have developed thus, without its natural transition from traditional filmmaking [11].

3 Language, Techniques and Hardware

3.1 Virtual Reality

In the most basic of explanations, virtual reality is a way to immerse users in an entirely virtual world. In VR, a user will, almost always, wear a headset which encapsulates the entirety of his/her viewing perspective. In other words, what you see is completely dictated by what is seen in the headset. In more elaborate setups, additional senses are utilized. In VR arcades, guests are often strapped into contraptions which simulate movement, sound is customized, air may be blown on the subjects to mimic the feelings our brains expect in certain situations, and some sort of handheld tools are used to allow user interaction within the environment.

3.2 360-Degree Video

360-degree video is often lumped into the VR category, but the differences are fairly finite. The mistake commonly happens based off of the ability to watch 360-degree videos in a virtual reality headset. In fact, a 360-degree video can be viewed in many ways. You might watch one on your phone as you scroll through your Facebook feed, using your finger to move around to get the 360-degree perspective. Or, perhaps perusing YouTube videos and using a mouse to gain the 360 view. And, maybe you really are wearing a VR headset to feel fully-immersed in the video playing.

However, aside from sound, you are missing the other senses, and you are lacking true freedom. Without being able to move through the 360-degree environment at will, you are not being given the six degrees of freedom required for VR. Sure, you might be travelling along as a bystander by way of a 360-degree GoPro camera strapped to a biker's handlebars, but you do not get to control the movement. What's more, you are not allowed to interact with the environment.

In 360-degree videos, the content creator determines the narrative of the video. It progresses from scene to scene, allowing the curator to tell the story they want to tell and present a very specific narrative to the audience. Virtual reality, by virtue of its relative freedom of motion, gives control of this narrative to the user. While the creator can present a narrative, the way in which it is presented is up to the user and the way they interact with the virtual world. This allows the content and the brand utilizing it to gain much deeper engagement as the user is allowed to create their own experience [13].

3.3 Degrees of Freedom

Degrees of freedom refers to the amount of flexibility a user is given within an environment. In our current example, the X axis represents backward and forward. The Y axis is left and right, and the Z axis is up and down. Where the three lines intersect is the point of origin.

Combining the three axises and their circular representations gives us a 360-degree compilation. While you have the option to look around in a 360-degree spherical world, and there is a certain level of freedom associated with that (after all, you can't do that when watching a normal cinematic movie), this is not truly what is intended when referring to degrees of freedom.

Instead, DOF intends to determine whether or not you can move or manipulate things on those same axises. Can you jump backward and forward? Can you slide from left to, right? Are you able to walk up and down stairs? Since you can do these things in real life, virtual reality, if it's to be truly virtual, must also allow for these six degrees of freedom.

3.4 A 360 Project

Education has always been an important factor of life, being continuously analysed, in the attempt to improve its delivery in today's classrooms. Although much has been done to give education more interesting ways of delivery, yet there are several generic instances when educational techniques used in today's classrooms are deemed as outdated, both by educators and their students [14].

Subjects, which are meant to enhance the knowledge and appreciation of a culture's heritage, can at times not be exposed to students in the most exciting way possible, so as to enhance learning and maximise understanding. A country's heritage is the map to its history. The accumulation of its languages, including its artistic endeavours and representations, stand as a reminder of our ancestors who have toiled hard to create the story that we are nowadays striving to keep alive and further enrich through contemporary means [15].

Technology has become a tool which stands alongside the brushes and rasps of the artists and sculptors of antiquity. Today, computers and their burgeoning peripherals have given art newer twists and further methods of expression, which can in turn augment the way students are drawn into the enchanting world of their country's heritage.

This project involves the creation of a 360-degree film environment, which combined with the enrapturing use of spatial sound [16] will recreate an epic moment in the fairly unknown initial stages of the Great Siege of Malta. This technology, through the artistic use of appropriate editing techniques, usually attributable to traditional film, will essay to incorporate new and developing 360-degree film language with traditional film language to further increase the element of immersion in the story depicting this salient point in history. This immersion is aimed not only to excite the young minds of students through the narrative techniques used, but further create compassion through an increased sense of empathy [17]. Several authors have mentioned empathy in tandem with 360-degree film and Virtual Reality. Many artists have gone the whole mile and delivered haunting experiences of such empathetic moments, through these same technologies. This project aims to use such authors' and artists' works together with an appositely filmed 360-degree experience, both visual and audio [18] to discuss the increased empathetic level of such experiences [19], that will in turn create the appropriate factors needed to increase the knowledge acquisition rate of students in respect of their specific heritage.

Janet Murray, in her chapter on immersion states,

> We seed the same feeling from a psychologically immersive experience that we do from a plunge in the ocean or swimming pool: the sensation of being surrounded by a completely other reality, as different as water is from air, that takes over all of our attention, our whole perceptual apparatus [20].

mentioning perception again as an apparatus in comparison with the term immersion. What is this element that increases perception and intensifies immersion? Chris Milk had hinted heavily at this element, through his travels from traditional film to VR technology [11]. Milk realised that it is the narrative which moves the individual to understand the story that he/she is experiencing, this same narrative that he carried forward from his films, to his art installations, to his interactive art to his VR productions [12]. The recurring element, that survives the passage of time and the transition of media is narrative, which creates, according to Merleau-Ponty [9] and Nisha Gupta [21] a "heightening of perception". Kevin Brooks [22], analyses immersion and hints at the reduction in the "heightening of perception". Brooks blames people's attention spans. Because of a new culture of what he calls "irrelevant information and cult of instant-gratification", a culture also reiterated upon in some detail by Simon Sinek in his Ted Talk, "Why Leaders Eat Last" [23]). Brooks determines this "heightening of perception", hence immersion as,

> something [that] can be seen. Heard and manipulated, if not also felt [22].

But Keven Brooks does not distinguish between traditional film and VR,

> delivering the sights and sounds by immersive media can cause a strong visceral and cognitive belief in what is experienced. It is belief that puts the real into the virtual reality [22].

The development of such immersive media, has existed since storytellers started providing this believable immersive experience since the beginning of humanity and have developed through the years into countless forms.

Similarly, Claude Bremond had specified that,

> [Story] is independent of the techniques that bear it along. It may be transposed from one to another medium without losing its essential properties: the subject of a story may serve as argument for a ballet, that of a novel can be transposed to stage or screen, one can recount in words a film to someone who has not seen it. These are words we read, images we see, gestures we decipher, but through them, it is a story that we follow; and it could be the same story [24].

Marie-Laure Ryan in her quest for the repositioning of the study of narrative on the transmedial and transdisciplinary track, recognises both Roland Barthes and Claude Bremond as the

> founding fathers [of] a field of study that transcends discipline and media [24].

The truth is, that 360-film has been swift in encroaching areas which were more experimental in nature and hence 360-film gave them a further potential to be exposed on a diverse medium from the traditional, yet when it comes to film, 360-filming is simply a guest at the moment. The directors who have courageously taken on the banner of 360-filming pioneers are taking baby steps into understanding this relatively new challenge, a challenge which is not small by any standards. It has taken traditional film over a hundred years to develop its narrative language, most probably it will not take 360-film that long, but the transition will definitely be one ripe with trials that will still take its own time to fully comprehend [25].

Nonny de le Pena [26], co-founder of Emblematic, frames the issue perfectly when she says,

> Virtual reality offers a whole different medium to tell stories that really connect people and create an empathic connection.

She continues by stating that,

> Humans have long been able to immerse themselves in other worlds, through oral story or novels, painting, photographs, television, cinema and pure imagination. The mind does not travel alone—the body most certainly comes along for the ride [26].

De la Pena is working on a new form of journalism which puts the audience inside the story. Emblematic Group is one of the world's leading producers of immersive VR content. De la Pena's latest production, "Project Syria" [27] (see Fig. 4), met with

> mostly racist and politically vitriolic reviews [28].

Fig. 4 Project Syria enabling further immersion is a new trend of journalism [27]

Using a voice over to put the viewer further inside the scene that is being presented, "Project Syria", has used today's technology to reconstruct photos and footage of actual situations occurring in war-torn Syria. According to De la Pena, watching photographs and videos of this ongoing situation is not enough to create empathy. The reconstruction of actual footage enabled viewers to be immersed further into the reality of this situation once the headset is donned [26].

"Hunger in LA" [26] (see Figs. 5 and 6) is another project constructed by Nonny de la Pena in her attempt to create a more immersive form of journalism. "Hunger in LA" is an attempt at making society aware of the hunger which pervaded the Los Angeles community. The reception by viewers of this experience was nothing that De la Pena had ever expected. A man who collapsed in a diabetic coma, during this VR experience, was the pivotal point, which brought empathy from viewers as most crouched down in an attempt to rescue an actually "non-existent character" [17]. Similarly, "Project Syria" had an incredible reception during its five-day run at the Victoria and Albert Museum, an empathetic reaction which surprised even curators at the museum for the incredible outpour of notes and comment in the guest book present at the experience [27].

Presence is defined as the subjective experience of being in one place or environment, even when one is physically situated in another. The feeling of presence would determine the effectiveness of a virtual experience or not [29]. This feeling of presence, hence engagement can shift from unintended actions in audience interaction to deliberate ones, evoking a sense of control. Some works can intensify this sense, pushing the user into modes of exploration caused by uncertainty. Four interaction phases were identified: adaptation, learning, anticipation and deeper understanding (see Fig. 7).

Fig. 5 A viewer of "Hunger in LA" at the 2012 Sundance Film Festival reaches out to touch the man who has fallen to the ground [27]

1. **Adaptation:** Participants adapt to the changes in the environment; learning how to behave and how to set expectations, working with uncertainty. This phase often occurs from the unintended mode through to the deliberate mode.
2. **Learning:** Participants start developing an internal/mental model of what the system does, this also means that they develop (and change) expectations, emotions and behaviours, access memories and beliefs. In this phase the participant interprets exchanges, explores and experiments relationships between initiation and feedback from the system. The participant therefore develops expectations on how to initiate certain feedback and accumulates interpretations of exchanges. This phase can occur from deliberate mode to intended/in control mode [30].

Fig. 6 A still from "Hunger in LA," an immersive journalism piece about the effects of the economic downturn on hunger. A man waiting in line at a food bank collapses into a diabetic coma. Using virtual reality, de la Pena brings the viewer into the scene as a witness [27]

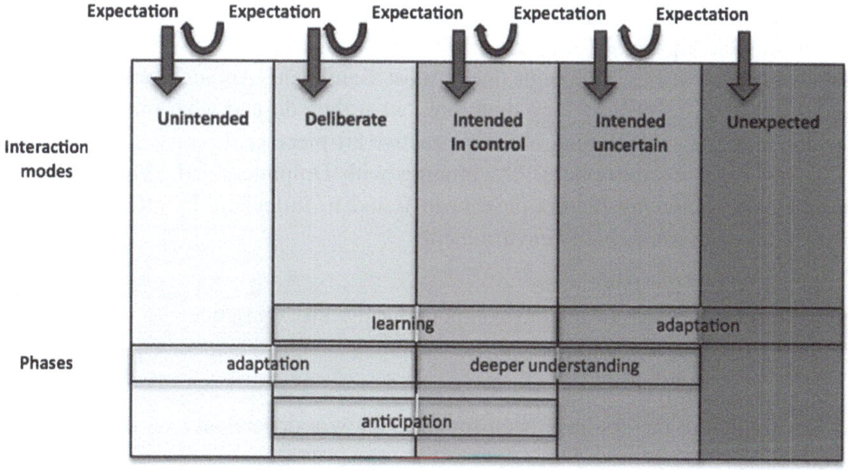

Fig. 7 Model of engagement: interaction modes and phases [30]

3. **Anticipation:** In this phase, participants know what the system will do in relation to initiation, in other words they predict the interaction. Intention is more grounded compared to the previous phases. This phase can occur from deliberate to intended/in control mode.
4. **Deeper understanding:** Participants reach a more complete understanding of the artwork and what his or her relationship is to the artwork. In this phase participants judge and evaluate at a higher, conceptual level. They may discover a new aspect of an artwork or an exchange not noticed before. This phase can occur from intended/in control mode to intended/uncertain mode.

Research has indicated that the learning self-efficacy, subjective norms, system quality, information quality and service quality all have a strong bearing on virtual reality learning, thus having a significant usefulness on perceived usefulness and perceived ease of use, while the perceived usefulness has a distinct, direct and positive impact on attitude of the users [31]. In this research users expressed the idea that virtual reality software system functionality is very important to them due to ease of integration into the educators' lives, creating opportunities of teaching in the future, yet can be difficult to learn to use [31]. From a varied group of subjects who used information technology on a regular basis in their lives and in their classes, it was noted that the users, educators in this case, expressed the idea that virtual reality systems provide complete teaching contexts and content and can provide reliable and immediate teaching, placing virtual reality use in class as a needed and trustworthy service [31].

Many studies show an ability in movies and imagery to elicit emotions. Human experience is defined as the content of direct observation or participation in an event [32]. The way that interactive art interacts with the audience is a crucial part of its being. The fulcrum of the art is what the work does rather than any other aspect. Thus, in this respect, the artist working on interactive art is quite different from a painter. Similar to interactive art, paintings, consider audience reaction through perception of colour relationships, scale, figurative references and so on. It is always the audience response to the work's behaviour that is most significant. Audience engagement is calculated in terms of how long they look, what they do and what interactions are developed in the experiencing of the interactive art-piece [33].

In two diverse experiences: "Swimming with Dolphins" and "Virtual Rollercoaster", the following factors have been found to influence the effectiveness of immersion in a virtual reality environment:

- Excitement of the experience.
- Comfort of peripherals and environment during the experience.
- Quality of the sound and images.
- Age.

The following factors have been found to show a dependence on the degree of immersion:

- Simulator sickness.
- Control.

- Excitement of the experience.
- Desire to repeat the experience [34].

The key issues for experience design, and for HCI in general, are the:

- Kind of experience
- Mode of engagement
- Phase of involvement.
- Viewpoint of evaluation [33].

The "Big three" structure of physical presence in virtual environments consists of perceptual, attentional and cognitive components [32]. All subjective experiences in such a built environment are quantifiable only by the person experiencing them [35]. John Dewey [36] states that almost all experiences arise from the interaction between a man and his environment, providing the possibility of understanding common patterns in various individual experiences. The analysis of these patterns in virtual reality environments is likely to expand knowledge concerning subjective experiences and interactive virtual experiences. Interaction is also acknowledged as one of the prime causes of presence in virtual environments [32].

Virtual Reality and 360-degree films are seen as a new, contemporary medium of storytelling. New rules are currently being established, a specific language for narrative being developed and new devices being created and explored [37]. While analysing all the aforementioned aspects, several characteristics need to be considered, namely the viewer, the plot, presence, theatrical application, guiding the viewer's attention and experimentation, which will depend very much on the tools being used to deliver the narrative to the viewer. Namely, experiences such as Oculus Story Studio, recommends that the viewer is given about 30 seconds to get used to the climate of the experience once he or she dons the tool being used to immerse oneself in the 3D environment [38]. This adaptation time gives the viewer a sense of acclimatisation, an opportunity to ease into this environment and learn how to react and conduct oneself in this new "world".

Eliminating CGI characters and ambiences, will remove the element of "uncanny valley" as described by Mori [39].

> The term comes from a graph created by Japanese roboticist Masahiro Mori that plots human empathy against the anthropomorphism of robots. On the graph, as robots become more realistic and we feel more and more empathy for them, the line trends upward. But as the robots' humanism approaches that of actual humans, our empathy for them — and the line on the graph — suddenly plummets [40].

This "Uncanny Valley" (see Fig. 8) plagues movies and video-games alike [41], so it is natural to assume that this will also occur in the world of virtual reality dooming the empathy that is considered of paramount importance for this project to succeed.

The plot is an important part of creating a believable experience. The plot does not need to be linear. Ian Palmer notes that the possibilities of non-linear story-telling has "exploded with VR" [6]. Palmer continues to say that this new format of narrative: Virtual Reality, is pushing story-telling to newer innovations, formats and paradigms

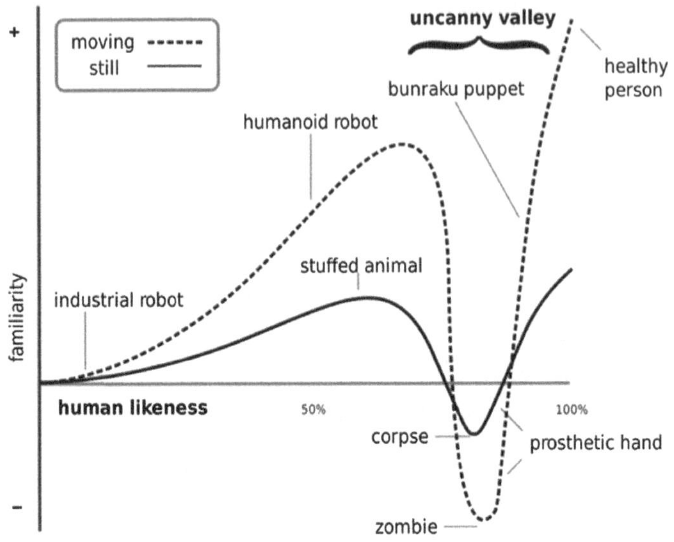

Fig. 8 Mori's 1970 graph, with reference points. He proposed that movement amplified the effect. "Familiarity" is used to mean "emotional response", and several semi-human concepts are listed as reference points. *Source* [39]

[6]. Palmer sees the skills needed as being fragmented across the wider industry at large, posing a gigantic feat for anyone trying to pull all the ingredients together.

The presence of the viewer is one of the most important attributers in a virtual world. Immersion is key in such cases.

> The present system of "transfiction" aims at extracting users out of reality when they enter the space of some camera. The captured image is analysed, the visual representation of people is automatically extracted and then integrated within a pre-existing story in order to construct mixed reality scenes. The users' attitudes and behaviours influence the narrative (interaction layer), with the explicit intent of making the immersion (of the user's image into the visual scene) a rich experience for all users [42].

It is imperative that the viewer is allowed to experiment and relates to the environment (the diegetic level) [42], yet there are possibilities that the viewer is not given full control of the virtual experience [25], but is led through the experience. As in film and theatre a semblance of control can be retained by the story-teller. As in Punchdrunk's "Sleep No More" [43] audience members are educated in the way they are to react with the actors and the scenery [44], yet in virtual reality there are no written rules, methods of conduct or acceptable forms of behaviour [38]. Yet Oculus Studio's "Henry" very adequately demonstrates to viewers what is acceptable or not and quickly puts viewers at ease with the experience that is unfolding in the virtual space. Story Study gives the viewers the time and chance to discover these nuances [38].

> At an incredibly emotional moment within Henry, he looks directly at you with sad eyes, acknowledging your presence within the scene [38].

further immersing the viewer into the narrative through the virtual experience itself.

PAVR's "Teleportaled", very uniquely deals with viewer integration by actually acknowledging the viewer's presence as an object present in the film itself [38].

> Teleportaled is a short movie filmed entirely in 360°. The actors discover an item, which is the viewer himself, and interact with the viewer throughout this virtual experience. Acknowledging the viewer in a direct way is a relatively new aspect of narration. Books, film and theatre transport the reader/viewer into the story through the immersive qualities of the narrative itself, a pre-scripted form of narrative [45].

Virtual reality gives the subjects in the experience the opportunity to related and acknowledge the viewer as that same viewer has been transposed into the realm of these subjects by donning the hardware being used to view the experience [45]. This keeps the viewer from becoming disengaged due to being forced into an outsider role as the narration is taking place [38].

Building the world wherein the narrative is about to occur also reserves paramount importance for an immersive experience. One aspect a director should consider is the "world-of-play", that is how narrative is created through the direction of audience attention [46]. Virtual reality is naturally immersive and thus is said to have more in common with theatre than with film [38]. As directors and narrators we need to ask ourselves questions as we progress through the filming of this project, questions like:

1. How are the characters to interact with the environment?
2. What kind of environment would lend itself best with the mood and type of story that is being told?
3. Where do the characters enter and exit the environment from?
4. What lighting would best aid the creation of the world that is being developed?
5. What sounds and sound effects would cause the audience to become more involved?

3.5 The Setting and the Story

All stories begin with words in your head; an event, an action that is yearning to be narrated. But, what is a story? This begs another question; Why are we talking about stories in a world of technological advancements in filming methodologies such as VR? The answer is quite simple really, everything starts with a story, everything happens because of a story, everything develops through a story. Humankind has been relying on stories since the first elementary communication methods were conceived. Stories are the element which keeps humankind in touch with its past and consequently, storytelling would be the art form which has developed over centuries of recounting events from family member to offspring. Let us delve deeper into the narrative uses and techniques needed to depict a good rendering of the story in 360 filming.

3.6 How Are the Characters to Interact with the Environment?

Primarily, the question arises, what is a character and why are characters needed? John Bucher [47] discussed 7 shards of narrative in his public talk in Malta in December 2017. One of these shards was adamantly stated to be "The Character". Bucher pointed out that even if the character is simply an inanimate object which is given human-like attributes, this would still be a character. He continued to add that there can never be a story without at least one character. But we will be discussing the functions of a character in storytelling and narratives in another article [47].

The title question points to two elements, existent in traditional filming, but taken to a new level of viewer expectation through the virtual ambience that is being discussed in this article. These fall within the narrative technique known as Point-of-View (POV), strongly present in both traditional and contemporary filming [48].

We have all seen, at some point in time, movies where the camera would take on an unexpected angle, and it is because of this angle, and particular way of capturing the shot, that the viewer is moved from a comfortable perspective to a more intense and, at times, uncomfortable way of observing what the director wants him/her to see. There are several POVs in traditional filming and these retain very similar traits when talking about filming for VR. Yet there are two particular POVs which take precedence over all other POVs whilst filming for any kind of VR experience.

The first POV is widely used in most VR filmed experiences. This would be the observer-within-the-action POV. The second POV goes deeper as it places the viewer not just within the action but as part of the action. Let's call this the direct-visual POV. What are the differences?

The first is the most usual depiction of a presence in filmed VR. It denotes the presence of a camera with two or more lenses. Discussing cameras and their evolution in recent years by a myriad of brands would mean another whole article by itself.

The advantage about cameras is that the choices are never-ending, the disadvantage is that many individuals shooting 360 video for the first time might make the mistake of going for high-end cameras immediately thinking that the better the camera, the better the results. In terms of resolution, frame-rate and stitching, this would definitely be the case, but otherwise, at times, keeping to more compact and practical camera solutions would be advisable, especially if the manpower is limited and the deadlines are tight. Back to POVs. This presence is usually a centered presence, and would not denote any movement from the first-person viewpoint. This scenario would simulate the viewer standing, crouching or sitting (depending on height of camera in relation to surroundings), in the midst of the subjects that/who are creating the ongoing action. Hence the viewer is more of an observer. This placement denotes a level of immersion which can tend to be partial as the viewer will not feel the need to interact with his/her surroundings. This method is used for situations involving the viewing of concerts, bands, interiors etc.

The second example of primary POVs in VR filming, is what we previously termed as the direct-visual POV. This takes into consideration the much-discussed difference between digital VR and film VR. The contention is always that film VR is technically 360 filming, and not, as is many times depicted, especially in 360-degree camera adverts, VR as seen in PSVR games. Digital enable users to interact with ease with their surroundings, due to their 3-dimensional (3D) build. In film VR this element is quite missing to date. The VR elements are limited to addons such as binaural sound [49], and direct interactions through "buttons" created in Unity. With these additions, and more on the horizon, 360-film can be given the elevated position retained, up till now, by digital VR experiences.

3.7 *Environments in 360 Storytelling*

The first clarification here would be to explain what I mean by "environment" in this particular context. The explanation would be twofold. The first type of environment being discussed, is the type of space that can either limit or increase the possibilities of interaction within a 360 film—this can be defined as the inclusion of presence due to available environmental possibilities, through available tech. The second type of space tackled in this article, is the space that is being filmed, the location itself. A good reference point, and starting point, for the first type of environment being discussed, is theme parks, and how much meaning is embedded in the this physical space. For instance, Disneyland uses a lot of forced perspective to make structures like Cinderella's Castle seem larger than it is and, by extension, for you to seem smaller and childlike. Yes, there is a manipulation of space. The space itself is designed to create a sense of nostalgia. Before anything even happens, the story has already begun, and you are immersed, without knowing, because immersion just happens and cannot be forced. The moment you understand and believe that you are in that environment by forgetting, even for an instance the real world, then the storytelling has functioned, the immersion is on.

3.8 *Space Is Dynamic*

"Defying the Nazis", takes you on a journey which is very similar to the theme parks mentioned earlier. But, the viewer is being transported, and although there is a semblance of movement, there is no element of dynamism present in this experience. In this PBS documentary by Ken Burns, memories of that voyage provide an emotional climax to the story of Martha and her husband Waitstill, two people whose incred-

Fig. 9 Using advanced sensing technology based on human stereo vision, the ZED® camera adds depth perception, positional tracking and 3D mapping to any application

ible acts of heroism had a monumental impact on society and history. This virtual reality experience recreates the voyage, putting the users right on the deck of the SS Excambion, creating a timelapse experience that takes them from a port in Lisbon to the Statue of Liberty in a matter of minutes.

The important thing to remember here is that set and setting are not just the background to the story or simply the vessel in which the action takes place. In order for it to feel dynamic, it has to be dynamic. It has to be engaged in the process. This is the power at your fingertips as an immersive storyteller.

Virtual reality is a way to immerse users in an entirely virtual world. In VR, a user will, almost always, wear a headset which encapsulates the entirety of his/her viewing perspective. In other words, what you see is completely dictated by what is seen in the headset. In more elaborate setups, additional senses are utilized. In VR arcades, guests are often strapped into contraptions which simulate movement, sound is customized, air may be blown on the subjects to mimic the feelings our brains expect in certain situations, and some sort of handheld tools are used to allow user interaction within the environment. Now, let's delve deeper in the first type of environment I mentioned earlier, the type of space that can either limit or increase the possibilities of interaction within a 360 film, the tech-related environment. Let's talk tech.

While the Samsung Round (Fig. 11) promises 3D video, the ZED (Figs. 9 and 10) and Intel's RealSense really manage to provide this 3-dimensional element so ethereal in 360-degree film shooting. Both technologies map surroundings and provide the depth which so many other 360-degree cameras lack. The Samsung Round, provides a similar depth, which is more of an illusion when compared to the other two. Considering that Samsung Round delivers 360-film recording and spatial sound together with the depth element, then this camera is very much currently seen as the complete package, which does not rely on other technologies to perform and create a deliverable.

It is the lack of 3D element in most 360-degree cameras, that is usually most felt by viewers during current 360-degree films. Although viewers can turn their heads and view subjects in motion and subsequently a narrative occurring, similar to a traditional movie, yet the picture still appears quite "flat". This detracts considerably from the "mood" and hence needs to be amplified by the "type" of environment that is being depicted. In short, diversity is key.

Placing the camera in the scene is key and important to storytelling [50]. To capture the action of the scene, you need to consider what role the person wearing the VR

Fig. 10 Real-time 3D mapping using ZED® stereo camera

Fig. 11 The Samsung 360 round® uses 17 camera lenses, eight stereo pairs positioned horizontally and one single lens positioned vertically on the device to livestream 4 K 3D video and spatial audio and create breathtaking 3D content with depth [49]

headset will take. Will the viewer be actually participating, or simply viewing the scene? This would prove various possible setups for the camera. What must always be kept in mind, in camera placement, is the height of the camera; as in, if the camera will be mounted low to the ground or hanging from a ceiling, or set up a few feet up from the ground to emulate a seated person, or standing person. The angle of the camera; as in, if the camera will be set up to emulate the view of an actual person, as if the viewer is peeking into an alternative reality. In any of these cases, the camera should always be placed in or around the center of the activity. If there is no actual activity, make sure to place items of interest which might help to "lead" the viewer towards the action, or if the viewer's gaze wanders, these items of interest would act as a magnet for the viewer's gaze back towards the action. Give the viewers freedom to explore as they wish, but always build the narrative in a way which acts as a guide, should the viewer get "lost" [51].

PAVR's Teleportaled [52], is an experience which takes the audience from environment to environment in the very short span of just over 4 min. In this film, it is the environment change itself which is of primary importance for the success of the experience. If the camera had filmed just one environment, there would have been very little of interest for the viewer. But, the camera's "jump" from one location to another break the monotony of the film and transposes the viewer into multiple settings without having to physically move an inch. Each setting is an escalation of interest and more for the viewer to view, hence the viewer is actually given a reason to move on through the experience without experiencing boredom.

In VR and 360-film, immersion is key. Technology has given the torch to VR in this case, but this does not detract the immersion possibilities for 360-video. In 360-film, it is the story-teller who creates the immersion. In traditional film, the 360-film elements were not present, yet immersion in certain movies was very strong. One would laugh, cry, boo, even feel hate for specific characters in a movie. That usually depended on the prowess of the story-teller. In 360-film, the same rules are being bridged ...and its working.

3.9 Characters and Environments in 360 Film

Another interesting factor that needs contemplation in the evolution of film-making using 360-degree cameras and binaural microphones is the entrance and exit of your subjects. Who are your subjects? Which subjects should be seen at a particular moment in time in the film? Where should your subjects stand within the 360 space? In which direction should they be moving, so that the viewer's gaze would intersect with their path? These questions have long been answered in traditional filming techniques, but applying them to 360-degree film is an entirely different story.

In traditional film, the process of leading the viewer into a shot is fairly straightforward, though not easy in any way, given that the director has a good sense of suspense, flow and a solid knowledge of narrative technique. Over the years of development of film as a medium of storytelling, through the works of major directors, film has bridged the gulf of immersion and empathy [19].

In the case of 360-degree filming, this situation postulates a very tangible problem indeed. The "where" in this case is the most important question one should be asking. Have you ever entered a room, a room that you have never been inside before, and immediately feel the urge to look around and acquaint yourself? VR and 360-filming both work on this innate premise within the human psyche [53]. "The Ring—VR Experience", a fan-made horror flick based on the famous franchise, "The Ring",

Fig. 12 The face of this Turkish soldier has been dirtied with makeup and actual wet soil for the filming of "First Contact". Even such a subtle change, will point towards a development in events, without actually having to explain it in complex and time-consuming ways

manages to do away with most of this room for error. It is very interesting how, although the viewer holds the control of where to look, the effects included in the editing are subliminal in passing on a covert message to the viewers to move the HMD around each time a glitch appears inside the viewers' filed-of-vision. Once the viewers turn their head, they will be met with a change in the scenario which should increase the scary element in this experience. Even though the viewers might not find the new subject immediately, they have been subliminally trained to look for something, hence the viewers would turn their heads around, knowing that something should be cropping up, until they eventually find it (see Fig. 12). This element of "training" your audience by inserting cues, immediately includes and increases the interactive measure in the experience. The viewer does not need to make use of any appositely programmed buttons, focal points, or any other external device, which would ultimately interfere with the immersive process, except for the innate, hence natural head-turning, that would interfere very little, if at all, with the immersive qualities intended [54].

So, in brief, plan the entrance of your subjects. Cue their entrance through foreshadowing. Foreshadowing is a highly effective plot device that makes the story plausible. It consists of an implication or introduction early in the movie of something that will be extremely relevant as the plot unfolds. In Ridley Scott's Thelma and Louise (1991), the gun that inflicts a wound in Act I is introduced much earlier, during the first ten minutes of the film. In Steven Spielberg's Jaws (1975), the foreshadowing is subtler, but equally worthy. Somewhere in Act II, Chief Brody (Roy Scheider) trips on a pressurized air tank, which he curses. In Act III, Brody shoves the same tank into the shark's mouth and makes it explode. In James Cameron's The Terminator (1984), an early scene shows the Terminator (Arnold Schwarzenegger) talking into a police radio imitating a dead policeman's voice, thus establishing this gimmick. Later in the film, the Terminator reproduces Sarah's (Linda Hamilton) mother's voice in order to discover Sarah's location.

Foreshadowing worked perfectly in traditional film. It is a technique which establishes elements which are of a strong importance at a later stage, sometimes even at a sequel stage. Foreshadowing brings forth the belief necessary, and that is desperately needed in a good movie to establish immersion. Once the viewer believes, the viewer is immersed. Loss of this belief, will jolt the viewer out of the narrative and hence immersion is lost [42]. In 360-film, foreshadowing is even more important than in traditional film. It is the technique which drags the viewer's eyes, and/or head around to where the action has been placed within the 360-degree environments. In traditional film narrative, foreshadowing is employed to explain an object, or a skill or talent, or a behaviour. In 360-film, all this still applies, while careful attention must be given to build more cues to enhance the movement guidance within the experience [55].

Such a technique can be very subtle or loud or brash. That, all depends on what kind of 360-film you are working upon. The dependence would remain wholly in the hands of the director and how he/she chooses to execute their narrative.

3.10 Lighting a 360 Film Setup

Lighting. A horror story in itself, without the complications of not being able to leave any red-heads, or yellow-heads, brollies, or box-lights running around. Even that XLR cable can and will be immediately visible within your sequence. It is a 360 experience, remember! My first advice is simple. Do not use extra lighting. Refrain from any lighting device. Stay with natural lighting. Photographers would understand this. What is better than natural flowing light!

One of the best options of lighting in 360 filming is definitely natural light (see Fig. 13). In the 360 project, "The Landing", great use was made of natural lighting and great care was taken to ensure picking a day with minimal wind, and a cloudy sky, which allowed for the perfect amount of light to filter through, which played well in the limited ISO setting on the Samsung Gear 360 camera being used.

But, having said this, is it totally advisable to stay away from lights? Absolutely not. In fact there are multiple ways to bring artificial lighting into your filming sequence, and still manage to hide the source, or not, depending on the setting in your narrative. Five elements that can be discussed are the following:

Fig. 13 Use of natural lighting in a 360 degree setup

3.10.1 Practicals

A simple lamp can have its bulb changed to emit powerful light and light up specific areas of the setting. These are the light fixtures that are already present within the filming setup. You can easily change the existent light bulbs within the fixtures for stronger lights and bathe the area in light, strong enough to light up specific areas of the setup. This can be used to create several types of ambience lighting needed to create a specific mood for your shot. These are considered the closest light source to natural light, i.e. sunlight.

3.10.2 Clean Plating

If practicals are proving to be insufficient, go for traditional light setups. Shoot with them in place and then remove them and shoot the scene again without them. This is called clean plating—the resulting two videos are overlaid in post and the clean plate is used to mask out the lights.

3.10.3 Exposure Plates

In 360 video you are often restricted to small cameras with low dynamic range or to scenes that have a higher dynamic range than you can capture perfectly across the whole 360 degree sphere. If the camera is in a locked position, you can shoot plates of the scene at different brightness to allow you to capture each window, light or the sky at it's perfect exposure. As with the clean plating, these can be overlaid in post to allow the best parts of the scene to be pulled through.

3.10.4 Hiding Lights

Lights can be hidden in a number of places in 360 shots, the simplest being on the tripod or monopod of the camera itself. Strips of LEDs can easily be edited out in post when the floor of the shot (nadir) [56] is replaced. Otherwise, objects in rooms can be used to occlude the lights, hiding them behind furnishings for example (Stuart 2017).

3.10.5 Carrying the Light

Carrying the light is an expression to explain the boosting of the light from a natural source. For example, if there is a window in the shot but it's not giving enough light to lift the face of the main character in the shot, then this can be augmented with another light. The light can then be removed in post, unless it's already occluded by objects in the scene. The importance here is that the light on the scene looks natural, and if the user looks back to the source they don't see a light but, for example, a light that's part of the room or a window with the daylight outside (Stuart 2017).

It is safe to say that the general rules of traditional video filming no longer apply when you can't place lighting gear just out of the shot. For 360-film, it's all about practical lighting and embracing ambient light.

4 Conclusion

Recent developments in VR filmmaking have delved deep into techniques which had been previously unknown in traditional filmmaking or had totally different uses due to completely diverse filming techniques with a completely dissimilar set of rules. In VR the viewer does not sit back as part of the audience, but is taken into a different reality, the moment he/she dons a Head Mounted Display. More like theatre, rather than film, this offers the possibility for the viewer to be inside the experience, simulating a member of the audience being placed on a stage in a theatre during a full production. The setback is that the viewer does not have any idea of what the script is about and thus will have to use his/her intuition to follow the action as best as he/she can. In this case VR filmmakers are racing against technology to create means of helping the viewer adapt to the experience better and follow the action in a more liquid manner through the use of the afore-mentioned techniques, which although overlapping with traditional filmmaking theories, are currently being used for a different set of purposes.

Gardner [51] comes up with a list of these techniques in her TEDx talk at Princeton University. The interesting thing about Gardner's talk is that it is echoed by Geffen [50], in another TEDx seminar, this time in London. Although Geffen worked entirely through the use of computer generated imagery (CGI), particularly since he was introduced to this medium by his mentors at Pixar Studios, in both cases, Gardner; a university student, studying narrative in new media and Geffen; a veteran of filmmaking, followed the same trends and realised the importance of the same techniques, in the development of successful narration to a viewer inside the experience itself, unlike traditional cinema, where the viewer had been consigned the role of an observer to the action.

VR technology and 360-degree filming, provide the possibility to be in another place. Stereoscopy or stereoscopics, is a technique for creating or enhancing the illusion of depth in an image by means of stereopsis for binocular vision [57]. Most stereoscopic methods present two offset images separately to the left and right eye of the viewer. These two-dimensional images are then combined in the brain to give the perception of 3D depth [57].

VR and 360-film not only provide the opportunity to feel present inside the experience itself but also offers the opportunity to tell stories in a way which immerses the viewer further inside the filmed experience, bringing forth more meaning from the story in the way the narrative is being tackled [30, 47, 58, 59].

References

1. H. Tufford, La fee aux choux (the cabbage fairy), 1896 (2010). https://www.youtube.com/watch?v=MTd7r0VkgnQ
2. H. Klinkem, L. Surkemper, International journal for digital art history (2015). http://www.dah-journal.org/
3. C. Bishop, *Artificial Hells: Participatory Art and the Politics of Spectatorship* (Verso Books, Brooklyn, 2012)
4. M. Heidegger, *The Question Concerning Technology. Basic Writings* (1993), pp. 311–341
5. B. Bolt, *Art Beyond Representation: The Performative Power of the Image* (IB Tauris, London, 2004)
6. Virtualworldsweb, How can vr filmmakers keep focus on the narrative? (2016) https://virtualworldsweb.wordpress.com/2016/10/31/how-can-vr-filmmakers-keep-focus-on-the-narrative/
7. Google, Tilt brush: painting from a new perspective (2016). https://www.youtube.com/watch?v=TckqNdrdbgk
8. M. Billinghurst, Vsmm 2016 keynote: using ar and vr to create empathic experiences (2016). http://www.slideshare.net/marknb00/vsmm-2016-keynote-using-ar-and-vr-to-create-empathic-experiences
9. M. Merleau-Ponty, *The Visible and the Invisible: Followed by Working Notes* (Northwestern University Press, Evanston, 1968)
10. M. Loiperdinger, B. Elzer, Lumière's arrival of the train: cinema's founding myth. Moving Image **4**(1), 89–118 (2004)
11. Chris Milk. Chris milk. http://milk.co/treachery

12. ChrisArtifact Milk, The treachery of sanctuary- chris milk - creators project 2012 (2012). https://www.youtube.com/watch?v=_2kZdl8hs_s
13. C. Morris, Shoot killer virtual reality films with these 5 cameras (2017). http://fortune.com/2017/04/03/virtual-reality-cameras-360-degree/
14. M. Roussou, The interplay between form, story, and history: the use of narrative in cultural and educational virtual reality, in *International Conference on Virtual Storytelling* (Springer, 2001), pp. 181–190
15. J.A. Goldstone, *Revolution and Rebellion in the Early Modern World* (University of California Press, Berkeley, 1991)
16. M. Lalwani, For vr to be truly immersive, it needs convincing sound to match (2015). https://www.engadget.com/2016/01/22/vr-needs-3d-audio/
17. L. Wispé, The distinction between sympathy and empathy: to call forth a concept, a word is needed. J. Pers. Soc. Psychol. **50**(2), 314 (1986)
18. D. Murphy, I. Pitt, Spatial sound enhancing virtual story telling, in *International Conference on Virtual Storytelling* (Springer, 2001), pp. 20–29
19. G. Riva, F. Mantovani, C.S. Capideville, A. Preziosa, F. Morganti, D. Villani, A. Gaggioli, C. Botella, M. Alcañiz, Affective interactions using virtual reality: the link between presence and emotions. CyberPsychology Behav. **10**(1), 45–56 (2007)
20. J.H. Murray, J.H. Murray, *Hamlet on the Holodeck: The Future of Narrative in Cyberspace* (MIT Press, Cambridge, 2017)
21. N. Gupta, The cinematic chiasm: evoking societal empathy through the phenomenological language of film. Janus Head J. Interdiscip. Stud. Lit. Cont. Philos. Phenomenol. Psychol. Arts **14**(2), 35–48 (2015)
22. K. Brooks, There is nothing virtual about immersion: narrative immersion for vr and other interfaces (2003). http://alumni.media.mit.edu/~brooks/storybiz/immersiveNotVirtual.pdf. Accessed May 2007
23. S. Sinek, Why leaders eat last (2016). https://99u.adobe.com/videos/20272/simon-sinek-why-leaders-eat-last
24. M.-L. Ryan, *Avatars of Story* (University of Minnesota Press, Minneapolis, 2006)
25. J. Anstey, D. Pape, D.J. Sandin, Building a vr narrative, in *Stereoscopic Displays and Virtual Reality Systems VII*, vol. 3957. (International Society for Optics and Photonics, 2000), pp. 370–380
26. N. de la Pena, About (2016). https://emblematicgroup.squarespace.com/about-us/
27. N. de la Pena, 'you become a witness' - nonny de la peña's immersive journalism (2015). http://ochre.is/inspiration/you-become-a-witness-nonny-de-la-pena-immersive-journalism/
28. B. Crecente, Project syria vr experience arrives on steam to racist, vitriolic reviews (2016). http://www.polygon.com/2016/11/4/13507754/project-syria-steam
29. B.G. Witmer, M.J. Singer, Measuring presence in virtual environments: a presence questionnaire. Presence **7**(3), 225–240 (1998)
30. Z. Bilda, E. Edmonds, L. Candy, Designing for creative engagement. Des. Stud. **29**(6), 525–540 (2008)
31. C.-Y. Chen, B.-Y. Shih, Y. Shih-Hsien, Disaster prevention and reduction for exploring teachers' technology acceptance using a virtual reality system and partial least squares techniques. Nat. Hazards **62**(3), 1217–1231 (2012)
32. J. Takatalo, G. Nyman, Leif Laaksonen, Components of human experience in virtual environments. Comput. Hum. Behav. **24**(1), 1–15 (2008)
33. E. Edmonds, The art of interaction. Digit. Creat. **21**(4), 257–264 (2010)
34. S. Bangay, L. Preston, An investigation into factors influencing immersion in interactive virtual reality environments. Stud. Health Technol. Inform. 43–51 (1998)
35. M.J. Schuemie, P. Van Der Straaten, M. Krijn, C.APG Van Der Mast. Research on presence in virtual reality: a survey. CyberPsychology Behav. **4**(2), 183–201 (2001)
36. J. Dewey, Art as experience (Minton, Balch & co., New York, 1934). DeweyArt as Experience1934

37. R. Aylett, S. Louchart, Towards a narrative theory of virtual reality. Virtual Reality **7**(1), 2–9 (2003)
38. Oculus VR, Oculus story studio (2016). https://developer3.oculus.com/documentation/audiosdk/0.10/concepts/audio-intro-mixing/
39. M. Mori, The uncanny valley: the original essay by Masahiro Mori (2012). http://spectrum.ieee.org/automaton/robotics/humanoids/the-uncanny-valley
40. N. Wolchover, Why cgi humans are creepy, and what scientists are doing about it (2011). http://www.livescience.com/16600-cgi-humans-creepy-scientists.html
41. S. Lay, Uncanny valley: why we find human-like robots and dolls so creepy | stephanie lay (2015). https://www.theguardian.com/commentisfree/2015/nov/13/robots-human-uncanny-valley
42. A. McMahan, Immersion, engagement and presence, in *The Video Game Theory Reader*, vol. 67 (2003), p. 86
43. Punchdrunk, Sleep no more (2016). http://www.punchdrunk.com/sleep-no-more/
44. A. Soloski, Sleep no more: from avant garde theatre to commercial blockbuster (2015). https://www.theguardian.com/stage/2015/mar/31/sleep-no-more-avant-garde-theatre-new-york
45. S. Louchart et al., Emergent narrative: towards a narrative theory of virtual reality. Ph.D. Thesis, University of Salford, 2007
46. N. Keenan, Theater for the future. http://theaterforthefuture.com/tag/world-of-the-play/
47. J. Bucher, How to shoot virtual reality: Seven basics to launch you toward virtual virtuosity (2016). http://www.moviemaker.com/archives/summer2016/how-to-shoot-virtual-reality-seven-basics/
48. D. Djaouti1&2, J.-P. Jessel, G. Methel, P. Molinier, Interactivity: storytelling or storywriting? A closer look on videogames and narrative
49. C.I. Cheng, G.H. Wakefield, Introduction to head-related transfer functions (hrtfs): Representations of hrtfs in time, frequency, and space, in *Audio Engineering Society Convention 107* (Audio Engineering Society, 1999)
50. A. Geffen, Storytelling in virtual reality | anthony geffen | tedxlondon (2016). https://www.youtube.com/watch?v=TOtTIWBrtJI
51. K. Gardner, Virtual reality storytelling | kate gardner | tedxprincetonu (2016). https://www.youtube.com/watch?v=7LNEkmcR4BM&t=2s
52. PAVR. Pavr (2015). https://michael-ashton-z8jn.squarespace.com/
53. F. Carroll, Engaging photorealistic VR: An aesthetic process of interaction. Ph.D. Thesis, Napier University, Edinburgh, Scotland, 2008
54. M.O. Riedl, R.M. Young, From linear story generation to branching story graphs. IEEE Comput. Graph. Appl. **26**(3), 23–31 (2006)
55. M.O. Riedl, V. Bulitko, Interactive narrative: an intelligent systems approach. Ai Magazine **34**(1), 67 (2013)
56. S. Pye, What is a nadir hole? (2015). https://www.linkedin.com/pulse/what-nadir-hole-sarah-pye-
57. M.S. Banks, J.C.A. Read, R.S. Allison, S.J. Watt, Stereoscopy and the human visual system. SMPTE Motion Imaging J. **121**(4), 24–43 (2012)
58. J. Bucher, *Storytelling for Virtual Reality: Methods and Principles for Crafting Immersive Narratives* (Focal Press, Waltham, 2017)
59. F. Carroll, M. Smyth, L. Dryden, Visual-narrative and virtual reality, in *In Book of Selected Readings: The International Association Of Visual Literacy (IVLA 2004) Jahoannesburg, South Africa* (Citeseer, 2004), p. 2004

Augmented Reality

I'm trying to free your mind, Neo. But I can only show you the door. You're the one that has to walk through it.
Morpheus, The Matrix

The Effectiveness of an Augmented Reality Guiding System in an Art Museum

Wilbert Tabone

Abstract Throughout the past decades, art museums have employed a number of digital technology aids in order to enhance the visitor's experience and memory retainment. Augmented reality has proven to be among the most successful applications in this regard as studies have shown it to be the most effective in eliciting the interest of the visitors and enhance their learning. Most of these augmented reality systems have been implemented on tablet computers or hand-held mobile devices. This chapter explores the effectiveness of such systems and presents an experiment conducted using an augmented reality system operating on a head mounted display. Gathered results and user evaluation indicated that the system offered an enhanced museum experience and improved the users' concentration on exhibition, their art appreciation skills and more importantly elicited the visitor with a renewed interest in the artworks.

1 Introduction

Traditional user experiences in art museums usually involve commentary panels affixed next to artworks with detailed information about the display. Unfortunately most people do not spend sufficient time examining these sources of information to extract their knowledge [4], which is why alternative techniques and technologies have been tried in museums throughout the years to enhance the user experience. Examples include the use of audio commentary and interactive multimedia panels amongst other kinds of technologies. Whilst these are usually deemed to be attractive to the visitor, they are not always located next to the exhibits. Such a situation creates a barrier between the visitor and the artwork, hindering the opportunity for close observation of artworks whilst also taking up most of the visiting time of the user. Attempts have been made to improve this situation with the introduction of audiovisual guides, interactive multimedia kiosks, radio frequency identification

W. Tabone (✉)
MUŻA - Malta National Community Art Museum, Valletta, Malta
e-mail: wilbert@wiltabone.com

D. Seychell and A. Dingli (eds.), *Rediscovering Heritage Through Technology*, Studies in Computational Intelligence 859,
https://doi.org/10.1007/978-3-030-36107-5_10

(RFID) and quick response (QR) code based mobile systems. However, it has been found that these merely alleviate part of the problem as such systems may still cause the visitor to place too much emphasis on the information presented in the guide device rather than the object being exhibited [4, 15]. Such a situation goes against the pursuit of the ideal in art appreciation and therefore, alternative solutions have been sought. One such alternative technology that has been grippling visitors' interest in the last few years is augmented reality (AR).

In fact the Gartner Hype Cycle (Fig. 1) from the time of writing indicates that AR is currently in the trough of disillusionment meaning that market-wise the hype has died down but the market is stabilising in preparation to the plateau of productivity, when mainstream adoption takes off. The curve indicates that AR shall reach this plateau in the next five to ten years.

An AR system seamlessly combines real (physical) objects with virtual (artificial) objects in a natural environment so that these appear together in real time [18]. By presenting this integrated layer of computer-generated information and the users' perception of the real world, the final goal of AR is to encourage and support user interaction with the world around them [3, 13, 14].

In recent years, various methods that apply AR techniques to art museums have been proposed and explored. Studies have found that AR offers the museum visitors an interesting, fun, immersive and challenging experience which enhances their participation and motivation due to the creation of a combined real and virtual environment [5]. This has been noted to enhance the interest of children and young adults

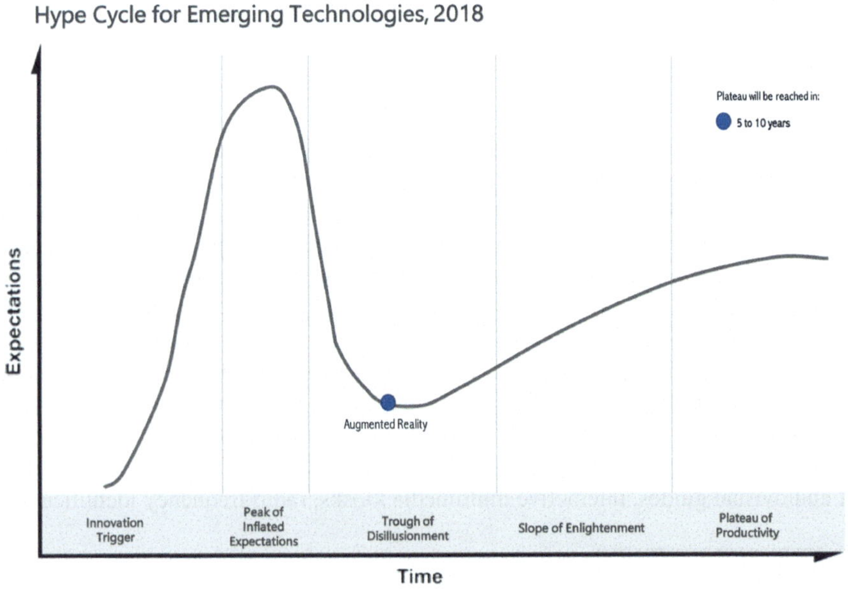

Fig. 1 Gartner Hype Cycle for emergent technologies (mid-2018). Notice how augmented reality is currently in the *through of disillusionment*. Adapted from http://www.gartner.com

in the arts [11] and improve the retention of information pertaining to the exhibited artworks [5]. In fact a study has shown that schoolchildren were more interested in repeating experiences when these involved AR [10]. Learning performance and interest of visitors has been known to improve when introduced to mixed reality systems in a cultural heritage context [6]. Furthermore, wearable AR technology has been found to have a positive effect on tourists' enjoyment when interacting with tourism attractions [17]. Such facts highlight the importance of providing visitors with innovative solutions for experiencing art, which is why these have increasingly been welcomed by the museum community [11].

Taking these findings into consideration we investigate the state of the art and previous implementations of AR in museums followed by a case study of our own at the Malta National Museum of Fine Arts in order to analyse the effectiveness of an AR system on the visitor's experience whilst also attempting to alleviate some of the problems encountered in literature.

2 The State of the Art

Augmented reality has been applied in a wide range of areas such as games, interactive films, education, training, marketing and navigation among others [5]. In recent years, augmented reality has also been applied in museums as an alternative guiding system. The main focus in this section will be on the previous systems or studies that are specifically tailored and tested in an art museum. These range from the contrasting of participant behaviour when exposed to AR-guided, audio-guided and non-guided tours [4], markerless systems [8], to experiments that test the subject's retention of a painting's appearance after exploring it using an AR system [9]. The following part of this section contains an overview of these systems which are predominantly implemented on handheld mobile devices such as smartphones, tablets or custom made portables.

2.1 Comparing Guiding Models

A museum guiding system, described as a self-catered guided tour was developed by [4] as a tablet computer experience containing a painting guide and a painting observation module. The first of these functions allowed the visitor to point the mobile device at a painting and a corresponding audio commentary guided them through observing the original artwork. The authors believed that in doing so, the users followed the four steps of art appreciation which are given as *the brief description of the artwork, the analysis of its techniques, an interpretation of the meanings in the work and judgment of values* [4].

Information is disseminated by the system through two layers, *A* and *B*, where the former is the description (textual) layer that is displayed in front of the pictorial layer (*B*). Layer *A* is split into four processes that correspond to the four-step methodology of art appreciation presented above (i.e.: description, analysis, interpretation and judgment). The first of these steps presents a simple objective statement to the user that describes what is in the painting and is then followed by the *analysis* step which assists the visitor in observing the artwork's design elements and principles, such as the proportion, colours, balance, shapes and lines. This additional layer of information is presented together with the ability of zooming in or out in the painting to observe the utilised painterly techniques. Through this methodology, users analysed the painting by comparing the information displayed on the tablet computer with the on-site painting, thereby still retaining the latter's importance. The *interpretation* step then uses zooming features to guide the users through various areas of the painting that contain important subtle messages or themes. This will aid the visitor to explore the significance and creative ideas of the painting which will likely lead to the user judging the work.

This kind of human-computer interaction (HCI) process simplifies the explanation process for features found in an artwork whilst simultaneously promoting an in-depth understanding and reflection of the artwork itself since it follows the art appreciation instruction [4]. The process also promotes a subjective psychological state of control, focus, attention and curiosity that leaves users with an enjoyable experience and a loss of self-consciousness and enhanced connection to the displayed exhibits. This phenomenon is defined by the authors as 'flow' and is triggered by the AR system which deepens the art appreciation skills of the users and places their visual focus on the exhibit.

2.1.1 Contrasting the AR-Guide with Traditional Guiding Methods

An experiment involving three groups, each assigned a different guiding mode (non-guided, audio-guided and AR-guided) was conducted in an art museum with visitor flow captured through cameras attached to the test subjects. Results indicated that overall the learning performance of the AR-guided group was superior to that of the other two groups. Similarly, the time spent observing a painting was highest for the AR-guided group followed by the audio-guided and non-guided group [4]. The AR-guided group felt that the guide device provided a greater motivation to learn and was more effective in delivering the message in a "livelier and less-boring" manner due to its extra visual commentary. In fact, the non-guided group expressed their desire to have a guide or commentary to aid them understand and appreciate the paintings as they felt that attempting to appreciate the artworks on their own proved to be ineffective. This concluded that the AR-guiding system eliminated certain limitations of other systems such as interactive media kiosks or video players which do not offer instant comparisons with what is being exhibited, leading to a reduction in the visitor's interest and viewing time.

2.2 Touchscreen Interface and Augmented Reality

A different approach was taken by [11] who presents a system whereby a static touchscreen is fixed next to the painting. An image of the painting, which can be zoomed in up to brushstroke level, is provided on the touchscreen together with further features that include social sharing and a 360° virtual tour of the painting. An accompanying tablet computer application was created to provide the AR features. Upon pointing the tablet device at the painting, an augmented version of the artwork is displayed to the user together with the relevant information on what is being shown.

In light of this system, [11] states that although some art curators still believe that the utilisation of technology will place art in the background, the authors argue that the artworks in this case become a resource and the museums act as the hosts in a kind of Internet of Culture. In such a way, the installations become a semantic network which users, especially young people, describe as interesting and positively influential in the elicitation of artistic interest (as proven by [4, 10]). In light of this, [11] warns that albeit still being informative, museum installations that do not introduce such technology and instead exclusively provide physical labels and panels to accompany the exhibited art are often deemed to be boring and unattractive by visitors.

2.3 Markerless System

A markerless AR system named *MuseumEye* was developed by [8] and tested at the Leeds and Egyptian museums. The authors argued in favour of markerless systems as it is cumbersome for users to follow the markers for each exhibit, and therefore removing that step from the user experience creates a faster and more focused tour. The AR operates in such a way that the real surface is considered as the origin point to the virtual object, so that the latter looks as if it has been placed in the real environment, hence gaining a realistic appearance. Augmentations and visual and haptic information are triggered based on an iBeacon triangulation process which tracks the user's location in the museum tour. The design of the tour was conceptualised as thematic tours, with audio content synchronised with the relevant artifacts in order to develop an ambient music that is appropriate to the museum context.

During the experiment, data was collected through direct user observation followed by a questionnaire. The respondent majority agreed that the developed application was useful for museum guidance and reported feeling immersed during the AR tour.

2.4 Visual Retention and Augmented Reality

An experiment aimed at analysing the subject's visual retention was conducted by [9] to discover if an AR-system enhances the user's ability to remember and objectify a painting after 24 h whilst using the gained information to solve new problems. Testing was conducted using a system which the authors named *AR-muse*. Exhibited artworks transformed into animated scenes when scanned with the system. Hence, the system operated with a magic-lens metaphor as it transformed exhibited artworks into animated scenes when the user scanned them using the tablet computer as a handheld lens. Users were also able to touch the augmented painting and pause the animations at any time.

Before each test session, the users were asked for a measurement of their perceived experience in art appreciation. The subjects were then exposed to either the artwork together with a text title and description, or the text together with the AR-system, or the painting with the AR-system only. Following the exhibition, the subjects returned 24 h later to answer questions pertaining to the different paintings that they were exposed to. The authors focused on various experiment variables including the number of paintings remembered, the recollection of their appearance by the subjects, the ability for the subject to objectify and therefore be able to create connections between works of art, the ability for the subject to question the artwork and the identification of the theme by the user across the whole exhibit.

In their results, [9] reported that overall there was a significant increase in the total dwelling time of the visitors when using the AR-system—a clear indicator that the system increased the subject's voluntary exposure to the paintings. It was also noted that the subject's ability to objectify was increased when AR was utilised, however on the other hand, AR was reported to be detrimental when coupled with text descriptions although evidence suggests that this still enhanced the objectification ability when compared to the text-description only group. An unexpected benefit from this 'distraction' is that subjects were made curious about the exhibition and the technology used and in fact users were observed examining the system and trying out new interactions to see the outcome. Although these interactions did not have anything to do with the content of the paintings, it is stated by [9] that such an additional stimulant to the experience could still have contributed to the enhancement of the subjects' objectification and learning ability.

Overall, [9] deemed their hypotheses satisfied and recommend that elements in the exhibition that interfere with the AR-system (e.g.: textual descriptions on panels) should be eliminated. Most importantly however, it was discovered that even if visitors had prior experience or training in art, the utilisation of AR would still bear unexpected benefits. This includes the ability to view 'paintings within paintings', which allows the users to gain a new perspective on the artwork, or an elicitation of further curiosity and interest in both AR technology and art. This newfound interest would have the potential of being a great benefit to the learning and analytical skills of the museum visitor and definitely to art museums all over the world looking to enhance their experience.

2.5 Limiting Effects of Augmented Reality Systems in a Museum Context

There are however certain disadvantages to AR-systems in art museums. One such disadvantage applies exclusively to AR operating on head mounted displays, where certain users may experience motion sickness, nausea and other symptoms after prolonged immersion [1]. A more general disadvantage, albeit subjective depending on the requirements, is that visitors utilising AR systems tend to be more alienated from other visitors during the experience. In fact, [4] reports that the group using the AR system did not discuss the paintings with their peers as opposed to the other groups. The AR-group focused intently on the paintings and their commentary due to the detailed presentation of the information and this may have been enough for some users and thus they did not feel the need to discuss the paintings further with one another. Furthermore, content in the interface may sometimes prove to be overbearing on the user due the number of facts they have to digest. Users from the experiment in [4] reported that at times the large number of words provided by the system did not let them digest or understand the most important points at once even though they had read or listened to the provided commentary.

On the other hand however, the majority of participants from the markerless AR experiment [8] had agreed that the technology actually encouraged them to interact with others. These contrasting findings raise the argument on whether user interaction is triggered by the exhibits themselves or the process of the learning experience. As previously noted, there are also moments were users interact to discuss not only the exhibits but also the technology used [8, 9]. It would be interesting to see whether such interactions would continue or eventually subside if AR systems were to become more common in museum settings.

Despite these less advantageous aspects, AR-systems are deemed to facilitate the development of art appreciation through the imprinting of paintings in the user's memory, therefore supporting the coupling between the AR system, the painting and the visitors irrespective of alienation [8, 9, 16, 19].

2.6 Design Considerations

A report by [5] provides a number of considerations to be taken into account during the design phase of museum interactive systems, particularly those that involve AR. The authors recommend that direct contact between the users and hardware such as keyboards, mice and touch screens should be minimised as much as possible. The system should also be user friendly and intuitive by providing a simplistic interface which is adaptable in such a way that it would be able to handle different but similar operations with no need for drastic modifications. Moreover, it is recommended that readily available hardware is utilised for cost efficiency whilst a simple and robust setup process should be created to reduce the risk of having a system failure [4, 5]. It is also suggested that in cases where handheld experiences are created, lightweight tablets or mobile phone sized devices would be more adequate, as users have been known to complain about tablet computers due to their bulky size [4] which created an inconvenience whilst operating the system and negatively effected the immersion process.

In respect to the different ways AR can be experienced, [7] notes how mobile augmented reality (MAR) is more widespread than head mounted displays (HMD) due to the fact that the former is more portable and the latter is sometimes criticised for limiting the field of view and comfort of the user. However, on the other hand, certain MAR systems are triggered using markers placed next to the artworks (e.g.: QR-codes) such that the system becomes obtrusive to the user's focus, whilst HMD are not operated in this manner. The takeaway message here is that using traditional marker systems is not recommended [7, 8]. Such a problem is alleviated by the aforementioned markerless system [8] and by a recently built projection-based AR system [12] which allows visitors to seamlessly switch between virtual and real representations of the exhibition [2] as they observe. Results indicated that whilst users felt positive about the experience, they reported finding such a system to be less immersive then HMD [12].

3 Malta Fine Arts Museum: Case Study

In order to analyse the effectiveness of such systems in an art museum setting, an in situ study was conducted at the Malta Museum of Fine Arts (NMFA) a few months before it closed its doors.[1] This study entailed the development of an AR system to be utilised by museum visitors as they wandered around followed by an analysis about whether the system enhanced their visit in any way.

3.1 Concept and Design of the AR System

The main concept for this system was to provide an information overlay for artworks which the museum visitor observes. It can be thought of as a quasi-virtual museum system, yet viewed as an overlay on reality. Having the system provide the necessary information about a particular exhibit would bypass the need of presenting physical textual labels next to each display [9]. The system not only displayed textual information related to the painting itself, but also enhanced details that would make the visitor engage more with the content. This included a portrait or photograph of the artist in addition to their name, and also a *similar-face* system for portraits[2] which was based upon [16].

[1]The MFA has been replaced by The National Community Art Museum (MUŻA).

[2]The test set for the system were paintings.

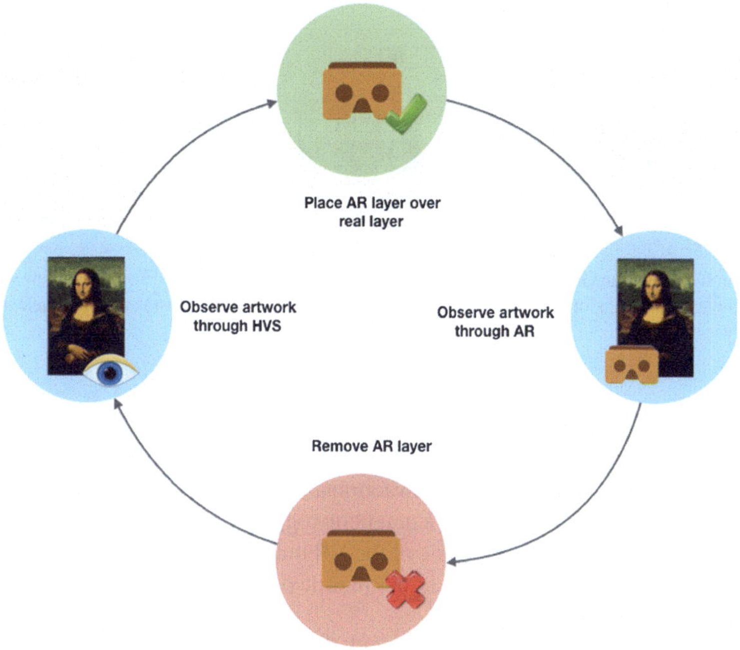

Fig. 2 Diagrammatic representation of the art appreciation circular journey

Curatorial direction advised that the art appreciation experience should be a circular journey starting and ending at the painting itself. This journey, represented in Fig. 2, starts with the museum visitor observing the artwork through their own human visual system (HVS); this is followed by observation after the placement of the augmented layer over the real layer. The user would then consume the information through the AR layer, and subsequently can remove this layer in order to look at the artwork again through their HVS, hence completing the circular journey.

In the previous section, the systems explored were predominantly MAR based. Whilst these were effective, they still posed a number of disadvantages, which were subsequently outlined. Hence, in the case of the system developed for the NMFA, a HMD approach was undertaken in order to explore their effectiveness when contrasted with MAR user experiences. In [4], the importance of the artwork was retained by having visitors use a tablet computer to allow them to compare between what they see on the tablet and the real artwork. The NMFA system still retained the importance of the artwork by allowing the user to see the real layer through the HMD at any time, as per the circular journey methodology.

For the scope of this experiment, the choice of HMD was the first generation Google Cardboard. Hence, in contrast to the systems discussed previously, the experimental AR system was built into a virtual reality (VR) headset. It was chosen due to its comfort, cost effectiveness, ease of access and most importantly it only requires a

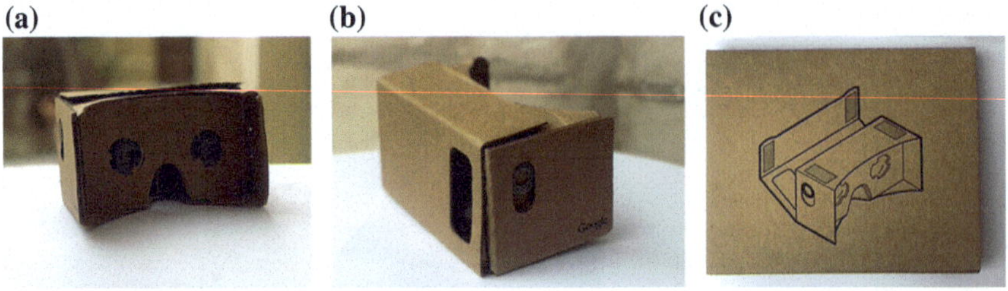

Fig. 3 First-generation Google Cardboard unit. **a** Back view with lenses. **b** Front-side view showing mobile device holding slot and the *select* magnetic button. **c** The unit comes as a flatpack for self-assembly

smartphone to be operated—making it available to billions of people. The Cardboard (Fig. 3), as its namesake suggests, is made of the thick paper material and comes as a flatpack (Fig. 3c) with assembly instructions. Other pieces include plastic lenses placed in the eye-holes (Fig. 3a) and a magnetic toggle switch called the *select button* (Fig. 3b) affixed to the right-hand side of the VR unit which can be pulled down with one finger to send a magnetic pulse to the mobile device in order to initiate an event on the respective program.

In order to satisfy the circular journey with the Google Cardboard, the background of the virtual world was set as the camera input of the mobile device. Hence, the artwork could still be viewed in the virtual environment as the information is overlaid. Moreover, the camera input fills the entirety of the background layer and moves with the user's field of view therefore giving users the ability to see their surroundings. This rectifies the issues reported in the previous section, where it was noted that HMD systems are sometimes criticised for limiting the field of view of the user [7].

The Cardboard interface contains a pointer system relative to the direct field of view of the user. Events can be triggered by aiming a yellow pointer at a *hotspot* (i.e. button or image) and pulling the physical magnetic (select) side button on the Cardboard. Users are familiarised to the pointer selection process and the use of the magnetic button through an instruction process which is displayed upon the first use of the system by the user. Prior to utilising the main interface, a small number of information frames (Fig. 4a) must be navigated by the user through pointing and clicking on the 'Continue' buttons in order to enhance their navigational skills. Once the last panel disappears, the main AR-system interface is shown together with an overlay containing a brief explanation of the provided buttons and navigation system (Fig. 4b).

A user initiates the augmented 'tour' by aiming the virtual pointer at the real painting (viewed through the camera) and pulling the select button. This re-centres the interface to place the painting view at the centre and displays a small high resolution copy of the painting as a virtual layer at the top edge of the viewport (Fig. 6a). A panel placed under the virtual painting contains its name and date of creation such that it is synonymous and metaphorically equivalent to the small labels included

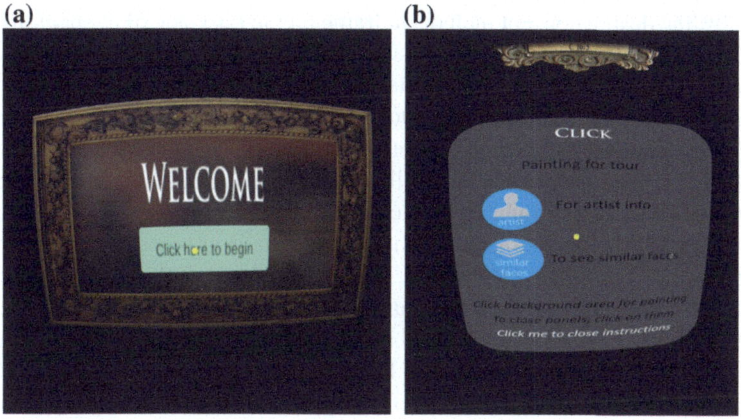

Fig. 4 Information panels preceeding main interface. **a** One of the introductory frames. **b** The instruction overlay explaining the various interface elements available to the user

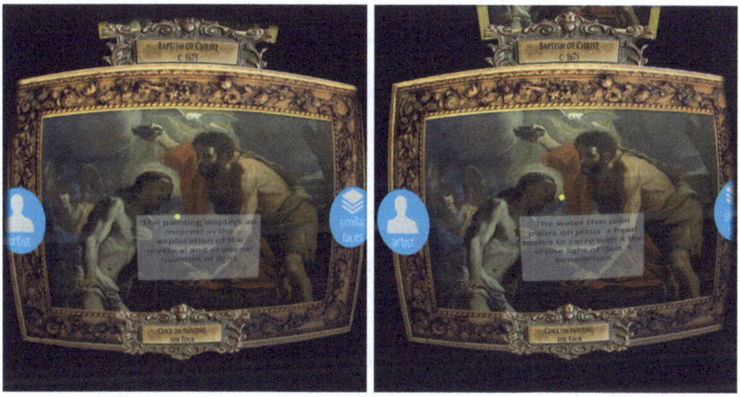

Fig. 5 Two consecutive points in the text tour, overlaid on the virtual painting. The yellow pointer can be seen on the white overlay

under paintings which display such essential information to the observer. Hence, the user can still view the real painting through the camera, whilst also being able to read the essential details.

A larger high resolution virtual version of the painting can be viewed through the Cardboard if the user selects and clicks on the small virtual copy. A further click on this high resolution virtual painting initiates a textual tour that is displayed as a number of small overlay panels on top of the scan (Fig. 5).

Each textual tour was designed to follow the art appreciation steps outlined in [4], by communicating information relating to the theme, colours and structures used in the exhibited piece. In order to reduce the number of selective clicks by the user, the information pertaining to these categories were grouped into one interface.

Originally, the design concept included a frame on the East side of the painting with detailed textual information. However, as [9] stated, such detailed descriptions both in physical form and in the AR-system itself are detrimental to the overall experience of the user. Therefore the panel was removed and its textual content summarised into points to be presented in the 'tour'. Curatorial direction was satisfied by this decision since it wished to do away with the small physical panels next to the paintings of the fine arts museums. This would allow the visitors to experience the artwork first and foremost without being distracted by the textual details, a process which was now made possible through the AR-system. The level of detail of the information in the textual tour was kept to the bare essentials since it has been found that it is the type of activity rather than the level of detail or the degrees of freedom in the interaction which governs the experience of users [10]. Furthermore, by utilising the text layer, the system conformed to the four steps of art appreciation [4] as enough information about the artwork, its message and technique were provided to the visitor to make an informed judgment.

In addition to the textual tour at the centre, two buttons were added on either side of the high-resolution painting frame in order to provide an unobstructed view of the high-resolution central virtual painting. The left is labeled as the *artist* button and the right labeled *similar faces* (Fig. 6b). When each button is clicked an overlay panel is displayed on top of the high-res painting. The artist overlay panel (Fig. 6c) contains their portrait or photograph, name and dates of birth and death. The user is able to select and click on the artist portrait/photograph to invoke a new layer containing a short biographical note (Fig. 6d).

In cases of portraits, the *similar faces* button triggers a new layer containing similar faces according to the various facial features present in the painting under examination, such that a *paintings within paintings* concept is included in the system. The similar faces displayed on the panel (Fig. 6e) are ranked and shown according to the similarity score achieved through the algorithm from [16]. The user is able to click on each face to see the source painting containing that visage (Fig. 6f). A further click on the source painting will change the entire AR interface to reflect the scenario of the user exploring that painting. This creates a link between each artwork in the museum. Hence, the system creates a network of artworks in such a manner as [16] intended in their work.

Once the information 'tour' is completed, the user would be able to return to the camera background through the various *close current interface* buttons which are represented by an encircled 'X' signifier. Following this, the user may wish to remove the Cardboard to look at the original painting—hence completing the circular journey.

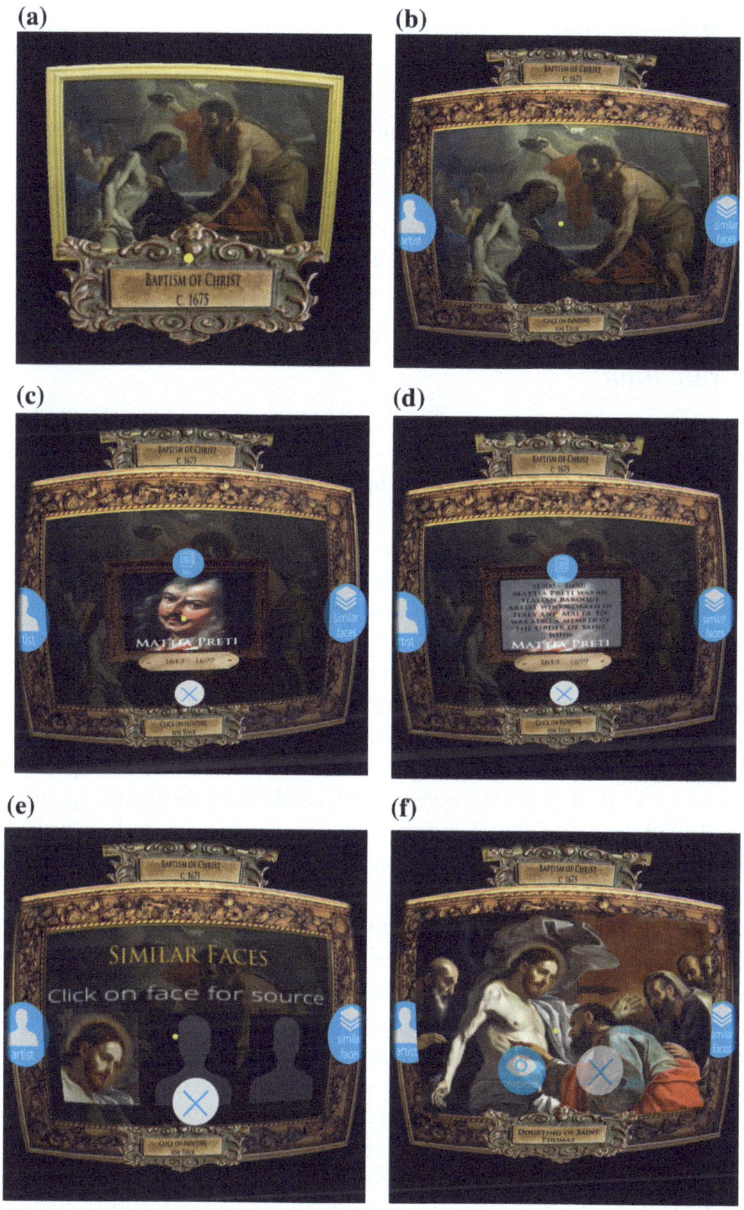

Fig. 6 The various information panels available in the AR environment. The dark background is the camera background in a dark room. **a** The small high-resolution virtual copy of the painting with name and date of creation. **b** Main interface of the AR environment, triggered after pulling the *select button* a second time. **c** Artist information interface triggered by selecting the *artist* button. **d** Selecting the *info* button from the previous interface triggers the artist bio overlay. **e** The *similar faces* interface triggered using the corresponding button on the main interface. **f** The source paining of a similar face, selecting the *explore* button marked with an eye signifier would display the information about that painting

3.2 Implementation

The experimental AR-system was created using the Unity 3D game-engine[3] in conjunction with the Google Cardboard Unity plugin.[4] This setup was preferred over the native cardboard SDK for Android and iOS due to Unity's powerful interface and the possibility of easily exporting the same project to a number of different platforms. In the case of the experiment conducted at the Malta Museum of Fine Arts, the AR-system was built for Android Marshmallow and run on an LG Nexus 5 device.

3.3 Evaluation

Evaluation of the system was done through a short questionnaire and an observational exercise. The observation phase entailed having the test subject wear the Cardboard and explore a chosen painting. Following a briefing explaining the overall system, the user was taken to observe a painting using the Cardboard. At the end of the observational phase, in which the user utilised all the interface interactions available, they were presented with the questionnaire in order to collect the necessary information for evaluation.

During the first part of the evaluation process, the user was observed using the system to flag any operational difficulties, such as problems in navigating the interface or problems related to the user 'feeling lost' when utilising the augmented interface. Qualitative data was collected during this process. Following the system test, a quantitative data gathering exercise was conducted utilising the questionnaire which collected data related to demographics, the art experience of the user and the user's evaluation of the system. Demographic information collected included the age group of the subject, their gender and nationality. Following this, the user was queried about their knowledge of AR-systems, how frequently they visit art museums and how artistic they deem themselves to be on a Likert scale of 1–5 (with the latter being the maximum). Another scale queried the respondent on how much they felt their art appreciation skills improved whilst using the AR-system when compared to the classical text-panel based setup. Two further questions asked whether the respondent prefers the AR-system's text tour over traditional text panels, followed by a rating of the user friendliness of the system and interface.

The effectiveness of the system on the learning process was also examined together with an analysis of how easy the displayed instructions were to follow and how user friendly the Google Cardboard was deemed. The latter part asked about the comfort of the Cardboard unit and the ease of interacting with the magnetic toggle button. Users were asked at the end on whether they would choose to utilise the system

---

[3]Unity 3D game-engine: https://unity3d.com/.

[4]At the time of writing, the Google Cardboard Unity plugin has been replaced by the Google VR SDK for Unity: https://developers.google.com/vr/.

again followed by an open question so that the respondent may freely write about any positive or negative aspects of the system.

3.4 Results

The evaluation phase was executed in situ on a typical week day. From the collected demographic information of 18 visitors, it was found that the majority of the respondents identified as male, belonged to the 13–18, 19–25 age groups and were European. The majority indicated that they had heard about either VR or AR before but they had never utilised such a system and rarely visited an art museum. The most frequent scores obtained on the *level of artistry* Likert scale were '3' and '4'.

Observation revealed a positive outlook by users in regards to the AR-system and its interface. The test subjects found the information layers easy to use and the interface aesthetically pleasing. Others commended the concept, stating that it is very interesting and important to adopt in museums in order to elicit the interest of younger generations. Further comments pointed out the fun factor the system introduces, which enhances the museum experience when compared to traditional linear tours where only text panels are available as sources of information.

Suggestions pertaining to the Google Cardboard device included increasing the depth of the device to cater for wearers of spectacles and alterations to the side button so that users of varying fingernail length could operate it with less difficulty. It is known to the author that the latter problem had been rectified in subsequent generations of the Google Cardboard, where the button was modified to enhance user friendliness. Most of the issues raised correlated with the observation of the test subjects during operation. Although the majority of the test subjects incurred no problems in utilising the system, some users had problems with the Cardboard toggle button as they were not sure how much pressure to apply, whilst others took some time to properly adjust to the point-and-click mechanism. Moreover, an elderly subject found the Cardboard uncomfortable to use for long periods of time whilst wearing spectacles.

The post-experiment questionnaire found that the majority of respondents (81.8%) believe that the system had improved their art appreciation skills. User friendliness of the system was given a good rating, with 45.5% of respondents awarding it a high score of '4'. Moreover, 81.8% indicated that they prefer the small text tour presented as an overlay on the high-res painting layer in contrast to physical text panels affixed to a wall next to the actual painting. Respondents also felt that they had learnt more about the artwork utilising the AR-system. Hence, the flow phenomenon from [4] was also present in this system since users felt that their concentration was completely focused on the artwork.

Respondents also believed that the instructions presented as an overlay (Fig. 4b) at the start of the AR experience were easy to follow as high '5' and '4' votes on the corresponding Likert scale were the most common. Similar results were achieved for the comfort and user-friendliness rating of wearing the Google Cardboard unit and using the physical button, with 63.6% of respondents indicating so. When asked if it was easy to find the required information layers (e.g. artist biography), the majority of respondents replied positively. Those who indicated otherwise stated that they preferred having the buttons situated in the middle rather than on each side of the frame.

Conclusively, the entire respondent body indicated that given the opportunity, they would utilise the system again.

3.5 *Further Suggestions*

If a museum chooses to utilise reusable HMD, it is suggested that the plastic versions of Google Cardboard are used instead, so that sanitisation prior to use is made possible. Alternatively, users may pay for the actual cardboard version at unit price as part of the admission fee and get to keep it for future use. Both the plastic and cardboard versions of the Cardboard may be branded, with the take home version acting as a souvenir as well.

Should the institution decide on using the Cardboard, then the users' phone may be used (requires installation of app) or the institution will have to invest in units which are provided to visitors during the duration of their visit. In such cases when a HMD unit requires a mobile device, it is imperative that the application can run smoothly and responsively on various mobile phone models—ranging from the higher-end models to the more economically-viable models. This can be ensured by using mobile-friendly assets or having a system which varies the quality of the assets and therefore the resource use is scaled according to the hardware specifications of the device.

At the time of writing, standalone virtual reality headsets are becoming more widely available and cheaper, removing the need of a third party mobile device to create the AR experience.

4 Discussion and Conclusion

Results obtained from the case study correlate with previously conducted research that present the benefits of utilising augmented reality systems in a museum environment. Judging by the participant's reactions and responses, the system had enhanced their visit at the Fine Arts Museum.

At the moment, the only barrier for introducing such HMD systems in museums is the current state of global market uptake of virtual reality headsets. Although VR has been the next big thing for over twenty years, mainstream uptake of the technology has yet to happen. On the other hand, AR on mobile devices is growing in popularity with the introduction of Apple's AR Kit, Google's AR Core and Unity integrating the VUFORIA augmented reality creation library directly into their engine. Hence, while at the moment museums are employing mostly AR systems which work on handheld mobile devices, there will be space for fully immersive AR experiences in the future as VR goggles (HMD) become more commonplace in a museum setting. The visitor is ready for it.

Furthermore, although the benefit of AR-guided systems such as the one presented in this chapter has been proven, it is still important to remember that for some visitors, new technology may be daunting or even alienating. Others may still prefer a traditional experience when visiting a museum. Hence, AR-guiding systems should be included as an option so that users can decide on whether to go on a traditional or enhanced journey.

In conclusion, analysing the results obtained in previous studies and including our own, it can be said that users have repeatedly confirmed that such systems aided their learning during the art appreciation exercise and enhanced their understanding. Such feedback is encouraging for introducing these systems widely in all art museums, creating further dialogue between visitors and incentivising them to visit again.

Acknowledgements The author would like to thank Dr. Sandro Debono who was the senior curator of The National Museum of Fine Arts for his insight during the course of the conducted experiment. A further thank you goes to all the staff at the museum for their feedback.

References

1. E.F. Anderson, L. McLoughlin, F. Liarokapis, C. Peters, P. Petridis, S. Freitas, Serious games in cultural heritage (2009)
2. O. Bimber, R. Raskar, *Spatial Augmented Reality: Merging Real and Virtual Worlds* (AK Peters/CRC Press, Boca Raton, 2005)
3. T. Caudell, W. Barfield, Basic concepts in wearable computers and augmented reality, *Fundamentals of Wearable Computers and Augmented Reality* (CRC Press, Boca Raton, 2001), pp. 19–42
4. K.-E. Chang, C.-T. Chang, H.-T. Hou, Y.-T. Sung, H.-L. Chao, C.-M. Lee, Development and behavioral pattern analysis of a mobile guide system with augmented reality for painting appreciation instruction in an art museum. Comput. Educ. **71**, 185–197 (2014)
5. C.-Y. Chen, B.R. Chang, P.-S. Huang, Multimedia augmented reality information system for museum guidance. Pers. Ubiquitous Comput. **18**(2), 315–322 (2014)
6. M. Christofi, C. Kyrlitsias, D. Michael-Grigoriou, Z. Anastasiadou, M. Michaelidou, I. Papamichael, K. Pieri, A tour in the archaeological site of Choirokoitia using virtual reality: a learning performance and interest generation assessment, in *Advances in Digital Cultural Heritage* (Springer, 2018), pp. 208–217
7. P. Coulton, R. Smith, E. Murphy, K. Pucihar, M. Lochrie, Designing mobile augmented reality art applications: addressing the views of the galleries and the artists (2014)
8. R. Hammady, M. Ma, A. Powell, User experience of markerless augmented reality applications in cultural heritage museums: 'museumeye' as a case study, in *International Conference on Augmented Reality, Virtual Reality and Computer Graphics* (Springer, 2018), pp. 349–369
9. W. Lu, L.-C. Nguyen, T.L. Chuah, E.Y.-L. Do, Effects of mobile AR-enabled interactions on retention and transfer for learning in art museum contexts, in *2014 IEEE International*

Symposium on Mixed and Augmented Reality-Media, Art, Social Science, Humanities and Design (ISMAR-MASH'D) (IEEE, 2014), pp. 3–11
10. D. Michael, N. Pelekanos, I. Chrysanthou, P. Zaharias, L.L. Hadjigavriel, Y. Chrysanthou, Comparative study of interactive systems in a museum, in *Euro-Mediterranean Conference* (Springer, 2010), pp. 250–261
11. R. Pierdicca, E. Frontoni, P. Zingaretti, M. Sturari, P. Clini, R. Quattrini, Advanced interaction with paintings by augmented reality and high resolution visualization: a real case exhibition, in *Augmented and Virtual Reality* (Springer, 2015), pp. 38–50
12. S. Schmidt, F. Steinicke, T. Huang, A. Dey, A projection-based augmented reality setup for blended museum experiences, in *ICAT-EGVE 2017-Posters and Demos* (The Eurographics Association, 2017)
13. T. Starner, Wearable computing, *Fundamentals of Wearable Computers and Augmented Reality* (2015), p. 13
14. T. Starner, S. Mann, B. Rhodes, J. Levine, J. Healey, D. Kirsch, R.W. Picard, A. Pentland, Augmented reality through wearable computing. Presence: Teleoperators Virtual Environ. **6**(4), 386–398 (1997)
15. Y.-T. Sung, K.-E. Chang, H.-T. Hou, P.-F. Chen, Designing an electronic guidebook for learning engagement in a museum of history. Comput. Hum. Behav. **26**(1), 74–83 (2010)
16. W. Tabone, D. Seychell, Recognising familiar facial features in paintings belonging to separate domains, in *Computer Analysis of Images and Patterns* (Springer, 2015), pp. 125–136
17. I.P. Tussyadiah, T.H. Jung, M.C. tom Dieck, Embodiment of wearable augmented reality technology in tourism experiences. J. Travel Res. **57**(5), 597–611 (2018)
18. D. Van Krevelen, R. Poelman, A survey of augmented reality technologies, applications and limitations. Int. J. Virtual Real. **9**(2), 1 (2010)
19. S.A. Yoon, J. Wang, Making the invisible visible in science museums through augmented reality devices. TechTrends **58**(1), 49–55 (2014)

Augmented Reality in Cultural Heritage: Designing for Mobile AR User Experiences

Vanessa Camilleri

Abstract Cultural institutions are increasingly manifesting their need to become more relevant in a society which is more technology driven than ever before. Today's generations seem to be present more in the online sphere, choosing Google for virtual research over physical spaces. This is making more institutions feel at greater risk of becoming more irrelevant as human presence has been markedly declining. In this chapter, we describe Mobile Augmented Reality (mAR), as one of the ways which cultural institutions are choosing to adopt to attract a bigger and more diverse audience in an attempt to bring people closer to cultural heritage and art.

1 Introduction

Cultural institutions are increasingly manifesting their need to become more relevant in a society which is more technology driven than ever before. The Knight Foundation, a foundation based in USA that supports community engagement through information, has recently funded a number of initiatives helping museums integrate technology for more user-centric experiences. Chris Barr, the director of arts and technology innovation at the Knight Foundation [11], reports that although technology may present itself as a rather de-humanised experience in a context such as cultural heritage, there are various aspects of it which can help users empathise, and connect with past and historical experiences in a rich, and emotional way. People visit museums and sites to be able to relive a chunk of history and learn more about their origins and their identities. However, in this digital era where people are connected with objects, information and other people through the online medium, the mere transfer of information through the use of a guide or visual aids, may not be enough to engage users and provide an enriched experience. The crucial scope of the digitalisation of culture artefacts is to help humans connect to the story being told or presented over the technology-powered medium.

V. Camilleri (✉)
University of Malta, Msida, Malta
e-mail: vanessa.camilleri@um.edu.mt

D. Seychell and A. Dingli (eds.), *Rediscovering Heritage Through Technology*, Studies in Computational Intelligence 859,
https://doi.org/10.1007/978-3-030-36107-5_11

This notion of using technology to facilitate an emotional and affective experience in users is not a new concept. Chris Milk, founder and CEO of Within (Virtual Reality) [32] has been increasingly advocating the use of virtual reality (VR) and immersive spaces to foster and support empathy. Milk has been involved in immersive installations with some of the most visited museums including the Museum of Modern Art (MOMA) in New York, and the Tate Modern and Barbican in London, amongst others. One of the greatest assets of these audio-visual media technologies for art and heritage, is the ability to use sound and pictures to create an illusion of space, and a sense of being which helps an audience become immersed in the surroundings. However, this is not the only benefit which such technology affords. The sense of being, augments its affective value creating a richer, more engaging and more memorable overall experience.

The dissemination of cultural heritage involves the ability of museums and institutions to transmit in people emotions related to a period in history placed into context [2]. In this way, visiting audience would be in a better position to understand and engage more with the exhibit, irrespective of whether the artefact belongs to tangible or intangible heritage.

The ability to feel as being in a different place, or a different time within a context, helps the audience achieve a heightened empathy with the situation. This sense of "being there", places the audience not just as an external observer of whatever is the focus of the heritage representation, but as an active part of that representation itself. Technology such as that driving Virtual Reality (VR) and Augmented Reality (AR) media, facilitates this type of immersive environment through the merge of physical and virtual interactions. Exploiting what these media offer may change the concept of art and heritage, from the more traditional where it may sometimes be viewed as belonging to those skilled few, to one where everyone can be participative.

What we do know is that when the concept of Web2.0 emerged back in the early 2000s, its fundamental driving force was that of eliciting crowd participation. The cry for success was for users to move from the passive to the active: from a culture where users just read what was presented to them to one where they were provided with the right tools to co-construct information. This notion was at the roots of social media and from there on grew the concept of a participatory culture.

However, when we consider that for many cultural institutions, this concept is still rather alien then we start seeing a gap between what modern society has become accustomed to and what many cultural institutions are failing to provide. This does not necessarily mean that what cultural institutions provide is wrong—however many are opting for a more inclusive approach to bring art and culture closer to a society that is primarily not only more pro-digital but is also more participatory. Being part of a participatory culture is also about building shared narratives through co-creation and collaboration. Jenkins et al. [10] make reference to this notion of spreadable media, which goes beyond the mere use of technology infrastructure. In their work the authors illustrate how although the choice of technology and the relevant infrastructure is important to create the right setting of a participatory culture, the design of the user experience needs to be such as to facilitate the process, and support more users in creating meaningful experiences not just for themselves but to spread around their networks of friends and acquaintances. The implications of this for cultural heritage can not only be in terms of economic value but also for the social value in helping keeping heritage alive. It is also important to note that opening doors to a participatory culture is not a simple task to be taken lightly. Information can change, as it transfers from one person to the next within the network. It is a very important aspect to take into consideration at the design stage, in order to make allowances for personalisation of the content without compromising on the veracity of the original representation.

Some examples of museums that have decided to go for a more interactive digital approach include the Smithsonian Museum of Natural History [23], with their recent installation of Skin and Bones, whereby an augmented reality app allows users to view and interact with the skeletons from some of the oldest bone collections, viewing the different species as they were. Although the museum advises the audience to physically visit the exhibit hall, it has gone one step further to cater for those visitors who cannot travel. The online version provides the posters of the various exhibits, in such a way that people, can still view and use the app to interact with the bones, using their personal computers from their own homes. However, this sort of interactivity hasn't yet reached into the spreadable media domain—users making use of this technology, haven't yet reached the stage where they can personalise what they see and interact with and share it with their networks and peers.

However, at this moment in time we are seeing more and more museums and cultural institutions, exploit the pervasiveness of technology to bring culture and history closer to a society that is more pro-digital [2]. Using augmented reality to bring art or historical artefacts to life, is one of the ways that is being used to assuring more relevance in the digital culture era where the use of mobile devices, has become an almost prevalent practice in everyday settings and activities. This is markedly seen in public events, or community spaces where the individual's need to share publicly and online via social media specific moments in settings or landmarks, becomes more pronounced. When technology is heavy for the user to navigate through and access, then it becomes a major driver for the lack of uptake of that technology [29]. This is why the use of personal mobile devices that people have become used to, helps the case for AR to become one of the more popular technologies for cultural heritage. The next step in the future is that of bridging the gap between technology, cultural heritage institutions and people by adding the social value component to enrich the technology-driven experiences.

2 Augmented Reality

The origins of the word Augmented Reality (AR) stem from the words "to augment" or rather enhance real world contexts. In the case of the technology driving AR, it uses digital representations overlaid on the physical world objects, to enrich what people see, hear or feel [5]. The super-imposed computer generated images or animations over the users' view of the real world, is meant to change the users' perceptions

Fig. 1 Computer-generated image superimposed over real world objects in AR—photo taken from Virtual Reality Pop [30]

of what is seen through the phone's camera lens. A good AR design demands the harmonious merge between the real and the virtual world in a way that users do not feel the disconnection between the two worlds, but rather perceive them as a singular object—even if we observe different media [7]. For example, we might be seeing a computer animation, over a real-world object (Fig. 1).

With a good AR design, both virtual and real worlds can co-exist harmoniously, and this enriches the overall user experience.

AR applications vary in use depending on the different fields and domains in which they are applied. These applications can vary from a simple text notification displayed on a mobile phone to highlight different city locations for tourists, to a more complicated display of instructions during a surgical procedure. These text notifications may serve to emphasise certain features, or facilitate interpretation of the context, but more importantly this can serve to provide more accessible and timely information about the real-world object that is caught on camera. This makes the information highly relevant to the users' present actions [12].

2.1 AR Categories

There are four different categories of AR technologies that are adaptable to the context in which these are employed. These include marker-based, marker-less, projection-based, superimposition-based.

2.1.1 Marker-Based AR

This type of technology uses image recognition techniques to produce a result when the visual marker used is sensed by the reader (which can be the mobile phone camera). These applications will make use of the camera to capture the image, as the marker, from any other real world object. The simplest forms of markers are usually in the form of QR codes, which also makes them easy to recognise and do not require a heavy amount of processing power. Other types of markers may include distinguishable objects, with unambiguous patterns for easier identification. Calculated position and orientation serve to overlay the virtual object on the physical world in a congruous manner.

2.1.2 Marker-Less AR

Marker-less AR makes use of location-based or GPS technologies to provide data based on the location of the mobile device. The accelerometer embedded in smart phones provides the data needed in location coordinates for the app to trigger the AR based on a user's positioning. This finds its use in a number of marketing and tourism apps, using the phone's location to provide users with information about their nearby surroundings, such as places of interest, entertainment, etc.

2.1.3 Projection-Based AR

This type of AR technology does not require the use of mobile or wearable devices. It uses projected light by combining cameras to 3D sensing systems such as depth cameras in a way that it allows for digital displays to appear on any surface. This kind of AR can provide for a deeper and a more enriching interactive experience if the right environment conditions are met. Most often in these cases, any external or outdoor variables may affect the AR interaction.

2.1.4 Superimposition-Based AR

This technology uses object recognition techniques to partially or wholly replace an existing real object with an augmented view of it. This type of technology is also being exploited by consumer oriented retail businesses such as IKEA, that launched an app, so that its customers can view how furniture and other home goods can fit in their available space using their mobile devices.

According to Liarokopis [14], AR is characterised by 3 major drivers; it shows the merge between real-world and virtual objects, running in real time, whilst allowing for an interaction between users and virtual objects. We also believe that for increased mass and public use, there is another driver which needs to be taken into account. This relates to the ease of access and speed with which users can get their AR app going. The abundance of apps available for all operating systems, creates a new requirement for developers and providers to create an app that is easy to find, to use, and which engages the users actively (Fig. 2).

Fig. 2 IKEA retail app using superimposition AR technologies

2.2 *Mobile Augmented Reality Applications*

A field of AR applications which has warranted extensive research is the field of Mobile Augmented Reality (mAR). Kourouthanassis et al. [12] discuss a review of mAR design as applied to various domains arising from a number of studies amongst which is the emphasis on user experiences, design elements for content delivery, high fidelity versus low fidelity prototyping, and interface elements for enriched interaction. Most of the design-related research stems from the Human-Computer Interaction field (HCI), whereby interaction elements are integrated in the interface so that users are supported in the way they communicate with the mobile device and the surrounding environment [3]. Technology has transformed the way that people communicate—with others as well as with the various devices and online objects [22]. Kourouthanassis et al. [12] outline a number of broad principles for the design of mAR applications as applied to diverse fields and domains. We will expand on the ones which are most applicable for the field of cultural heritage and in particular with situated simulation AR experiences. These include: context provides content, task relevant content, content-privacy notification, provision of feedback and the procedural and semantic memory support.

Context provides content: In heritage institutions, context and environment become very important to provide the background to the digital information that is being overlaid. In this case, the mobile device itself will provide sensory input through the camera, accelerometer, GPS, gyroscope, etc. to create the context of use. The mAR space takes into account the dimensions of the object that is being displayed in the context of the entire real world captured via the camera.

Task relevant content: Content is filtered or personalised through adaptive interfaces based on camera and motion-detection mechanisms. This would enhance the overall usability of the app, providing a more engaging and personalised experience for the users.

Content privacy notification: Users are aware of any sort of personal information that might be collected during the experience. Users may choose not to disclose any personal information thus reducing any undue amount of confusion, or anxieties related to privacy.

Provision of feedback: Users are aware about the current state of the application, through real time feedback. This reduces frustration at not understanding what or how the system is behaving.

Procedural and semantic memory support: Familiar icons and consistent presentation ensures a more user-intuitive interface, thus reducing additional learning curves for using the app. This would ensure a more engaged audience, with increasing likelihood that users would return to experience the app.

(adapted from Kourouthanassis et al. [12] pg.12–17)

In summary, a well-designed mAR app experience would need to provide relevant content framed within a specific context that is in turn placed in the right physical environment. Adding to this, design elements need to cater for the right information, keeping an interface that is clutter free, yet informative and can be personalised by the user. Interactions need to be intuitive and take into account the device itself.

In addition to these principles for design, we wish to take into account another value that is fundamental for the success of a well-designed mAR app. This principle takes into account the social value through communication and sharing. In this regard, a mAR app would allow not only the taking of photos but it would encourage the sharing of and interaction with content. This form of sharing, is a major driver for the success of applications that are related to public spaces. This adds an affective quality to the app, that removes it from the isolation of physical space and sets it up as part of a larger frame within the community space. This would also enrich the user experience, rendering the application more usable and effective. Rogers and Sharp [22], describe the enhancement of sociability and the support of creativity as two of the desirable aspects of an effective and engaging user experience from an HCI perspective. This would, according to the authors, answer the question in relation to how long the users interact with the application, as well as how willing they would be to return to using the application. This would then ensure, that the application would not only be cognitively satisfying for the users, but it would also maintain its relevance for the target audience. In HCI there are a number of interactions which a user might have with the different devices, including communication through text, speech and gestures [22]. Interactions for mAR applications are modified slightly so as to include the manipulation of the rendered digital overlay, and the exploration of the digital space. In a situated simulation, such exploration has to occur in the context of the physical space and the environment that is being augmented [15].

3 The Case for Museums and Heritage Sites

Cultural Heritage is a very broad term used to define the legacy that generations of people have passed on through history. It is evidenced by the way the people live, and the way their values and traditions have been shaped. Therefore, cultural heritage, does not only refer to tangible artefacts, but it also includes the intangible aspect that people can gain access to, through narratives and audio-visual media [20].

In this chapter, we refer to the cultural heritage aspect that is tangible and which is often displayed using artefacts, objects or exhibits. Museums, and cultural heritage sites, are often the intermediate through which these tangible artefacts are exposed [21]. However, it is not always easy for people to physically visit a museum or an exhibition, due to geographical and time constraints. Most often museums are perceived as places where people walk in, take a look around and then walk out again. Exhibit halls may also disallow the use of mobile phones or devices to take pictures. This, coupled with the lack of interaction with the artefacts, as seen from the perspective of today's digital generations, may be a source of discouragement for the mass of the general public to visit cultural spaces. The problem then arises, on how to encourage people to visit museums and sites and on how to keep them engaged throughout. Although many tour guides do a great job of keeping their audience and visitors engaged, it is not always the case, especially with some museums and public exhibition displays, where guides are present on site to help the audience understand more about the artefacts. Technology, in particular AR can be used to enrich and engage the audience, but it can also be used together with hosts and guides, to augment the overall cultural experience [31].

Artivive [1], an AR tool designed for use by artists and museums to augment the art experience, has only been available since 2017. Yet during this time, the app has already accrued over 60,000 downloads, and people have spent a collective 1.3 million minutes engaging with art. During this past year, over 2000 artists in 65 countries have used the platform, supporting over 100 exhibitions, with 1.5 million scans of the artworks [4]. Although the numbers by themselves are impressive, it is worthier to note, that what this tool is doing is giving people the opportunity to augment the art experience through multi-sensory interactions via their personal mobile device. This interaction serves to augment the sensibilities of the audience to the artist's own interpretation of his artistic expression thus creating a dynamic collaboration between the artist and the audience [5]. This dynamic collaboration between artist and audience is not the only of its kind. The recent Egypt exhibition at the Royal Ontario Museum, Tomb of Kitines, has been enhanced using Holographic AR, in a way that 2nd century AD narratives about the Egyptian tomb can be passed on to the audience in a digital modality. In this app, museum visitors are able to enter inside the replica of the tomb, and wearing a see-through headset interact through gaze, touch, point, and grabbing motions [19]. Although there are a number of limitations and difficulties associated to the technology adopted, initial indications are that such interactivities can greatly enrich experiences in museums owing to the transformation of the narrative that can occur via immersive multi-sensorial approaches.

The sensitization to place and the context is also seen in Chang et al. [6], as they discuss the exploitation of AR to enhance users' cognitive and affective experiences to enrich heritage tourism. The authors maintain that AR not only enhances the experiences at an affective level, but that it can also be used so that users can associate better what they see to past events, and build new memories in the process. When a user feels as though he/she is living through an event or an occurrence, then there becomes a heightened sense of place and an illusion of being which helps the user to empathise more with the situation, and identified context. This in turn helps establish new memories based on the lived experiences as perceived by the brain. During their experiments with AR, Chang et al. [6] found that AR designed using the human-computer-context-interaction framework based on a historical geo-context embedded visiting guidance strategy improved the learning achievements by impacting the sense of place that the technology stimulates within the users. Although there are certain limitations to the assertions made by the authors what is interesting to note, is that the inclusion of AR technology to a heritage site, can effectively stimulate cognitive and affective skills, that can in turn be utilised to reach a deeper understanding of the site or exhibition that is being visited.

Adding on to this notion of aiding users construct new memories of history, Marques et al. [17] discuss how 3D visualisations of reconstructed historical sites or artefacts can help users identify with a different sense of place and being. 3D visualisation based on GIS and AR technologies could superimpose the digitally recreated places or objects on site thus giving the audience a better glimpse of historical contexts and events. The authors' work is based on an aqueduct in Lisbon that was declared as a national monument in 2010 and is considered as a landmark by many. They used 3D scanning and a 3D model, together with digital cartography based acquisition to provide the final AR of the original aqueduct. The importance of this work lies in the application of a marker-less application of the AR technology allowing users to scan the monument without the use of any QR code. The improvements in image recognition techniques are adding on to the advancements in AR applications in fields such as cultural heritage. Tom Dieck and Jung [26] discuss the added value for stakeholders of bringing technologies such as AR into cultural heritage. In their studies, they concluded that the integration of AR into museums, would bring added values, both to internal stakeholders (such as CEOs, managers, etc.) and external stakeholders (such as visitors, teachers, etc.). The issues which need to be addressed from an internal stakeholder perspective is that of overcoming the initial costs for developing, yet many are in agreement that the increased numbers of visitors to the museum, would justify the inclusion of technologies. External stakeholders in addition focus more on the historical as well as the social added value that technologies such as AR can bring to the museums. The added interactivity increases the experiential value of the museum exhibit or display, thus not only adding to the interest and engagement with the artefact, but also enhancing an affective connection to the object of the interaction [27]. This view of the added value as perceived by stakeholders is also supported by researchers [25]. Value creation, they argue, brings a competitive advantage that can attract more visitors to the site. However, the AR design needs to be sustained with a value creation framework that is contributed to by a number of stakeholders with diverse interests in the field. This framework depends on the contributions by the stakeholders during the requirements phase but is also dependent on a co-collaboration between developers, providers and users during the creation phase. This ensures a successful value-driven engagement with the intended audience.

4 Situated Simulations–Case Examples

Smordal et al. [24] discuss the concept of situated simulations as cultural experiences aiding learning. Knowledge building through context, or rather contextualised learning is not a new concept in itself. Lave and Wenger [13] had initially proposed a learning paradigm whereby learning is viewed as a social process, with knowledge that is co-constructed and situated in a context that would be meaningful for the learner. In the case of museums and cultural heritage institutions, we cannot but appreciate the educational value that can be held within tangible or intangible heritage. However, Smordal et al. [24] ascertain that this educational value of a site increases when context is attached to interaction in a socially guided environment. As a case study, they observe a group of students, working their way through Opera2222, a virtual 3D reconstruction of the Opera House in Oslo, with a surrounding setting with overlaid AR showing complete environmental destruction and decay. The scope of the project was that of understanding how situated simulations can help trigger communities of inquiry, that can lead to knowledge building. What emerged from the results was that the concept of situated simulation can be used for a broader scope than bridging the gap between schools and heritage sites.

Further to this, we took another example of situated simulation whereby Liestol [15] conducted research taking AR to another level by including a degree of storytelling, using situated simulations for digitally constructed historical re-enactments. In his work, he illustrates three episodes in history: the sinking of the heavy cruiser Blucher in the Oslo Fjord at the beginning of World War II, the ancient Panathenaia Procession on the Acropolis in Athens, and the D-Day landing on Omaha Beach in Normandy. One aspect which emerges from this research is that the storytelling aspect of the situated simulations becomes very important so that users can achieve a sense of place, time and space. Because the AR of the situated simulation is being viewed over the mobile device with a screen which is relatively smaller than the vision of the whole area, and which is also open to surrounding physical distractions, then the location chosen for the unfolding of the story becomes very important. Many historical episodes evolve over a period of time. Yet visitors don't always have hours, waiting to see history being re-enacted so what Liestol, worked on, was having the story told in a way that the user experience the re-enactment in a short span of time without causing incongruity and a break in the flow of the story. During the Blucher AR prototype, users were given more control over the level of interactivity by introducing the notion of a freeze-pane—the AR simulation is halted whilst the user is able to explore more the object of the AR (in this case the cruiser) in detail. The ancient Panathenaia Procession on the Acropolis in Athens, followed the same narrative pattern as the Blucher, however in this case the procession occurred in a limited geographical space, and historical interpretations about what used to happen during the procession and how varied. Once again the time-freeze option was included so that users could stop to read through the information layer, overlaid on the digitally AR animated sequence. The third historic re-enactment that Liestol discusses, is the D-Day landing on Omaha Beach in Normandy. This is a historical event that has been documented in great detail, and has seen dramatic unfolding of events on the beach. Taking on best practices from the previous two situated simulations, this AR app captures the assault on Omaha beach, compressing the entire duration of the battle, into a more audience-friendly experience. However, the difficulties which Liestøl discusses pertain more to the challenges associated with AR done in situ. As he argues, certain historical re-enactments, may have been previously presented in movies or tv—however few of the re-enactments would have been shot on the actual location. When designing for AR on site, then one has to take into consideration the real scale of the terrain and physical environment where the episode unfolds. Since most often locations and places are subject to changes in landscape and topographies over time, then when designing an app for AR-based situated simulations, the audience has to be informed in a way that doesn't break the flow, about how such actions can have been subject to change depending on the environment. When applying AR to situated simulations, interactivity plays an important role. An audience can be following the narrative, however if they don't have a degree of control over their experience, this will become another static audio-visual resource. Although AR technology may be innovative, there is a risk that unless there is some form of interaction, its novelty may wear off sooner than anticipated and the audience can quickly become disengaged.

Another important consideration in the design aspect of the AR app, is that the mixed reality is achieved by having the narrative told in full screen on the mobile device as the user views the real environment beyond the device. This choice for the full screen method seems to stem from the fact that the information displayed on the screen is continuously being updated as the user is moving from position—with a full screen display, the animations can thus be revealed without a break in the narrative. Breaks in the narrative are in fact an important focus of the design of the AR app. Building on theories of film and narrative from Genette [8], the author describes how the app for Omaha beach, uses techniques such as ellipses and summary where discourse is shortened from the actual time of the story as it unfolds, to flash over and condense the duration of the battle whilst still retaining a meaningful course. There are instances then when the narrative needs to take on the realistic duration of the scene where the discourse is set at the same pace of the scene, again in a way that provides more meaning to the user following the simulation.

5 SitCity—A Proposal Framework

At the University of Malta, we have experimented with a situated simulation to re-create an episode in World War II history that marked Malta.

In August 1942, Malta was a country ravaged by war. People gathered in the Capital City of Valletta were suffering from hunger, fatigue and sickness. The continuous assault of the ally ships by German and Italian bombers was rendering Malta inaccessible by air and by sea. Food reserves and medicine were depleted, and people were conscious that the end for them could be very near as they could either be hit by a bomb, or else die of hunger. That August, the allies launched Operation Pedestal—a convoy of ships carrying supplies that was meant to be the final effort before Malta would be forced to surrender to the Germans. The operation was a huge effort releasing merchant ships, battleships, aircraft carriers, and cruisers in an attempt to save the people on the island. On the 15th August 1942—a day which is still celebrated as a Public Holiday even today across the Maltese islands, the merchant ship Ohio made it into the Maltese Grand Harbour of Valletta, with oil supplies and food [16]. The atmosphere in Valletta according to [16], was of extreme jubilation as word got around and crowds of people lined up on the bastions to watch the embattled ship being towed slowly into the harbour. This episode has been recorded as one of the most memorable in the history of World War II in Malta due to the surge of courage, and motivation that the convoy brought with it and which helped the Maltese people to keep fighting and survive the war. For this reason, it was decided that for such an experiment the ideal set up would be in the same location as where this chunk of Maltese History unfolded, in the city of Valletta.

After much debate with a local historian on the best locations in Valletta where we could situate the AR app, we identified a site which is not only much visited by tourists but which is also one of the locations that has been documented as having been flooded by the cheering crowds when the SS OHIO was being towed into the Grand Harbour in the Capital City of Malta, Valletta. We also identified a target audience of foreign tourists as the major category more likely to use the AR experience due to the fact that the location was listed on the tourist map as one of the major landmarks in the capital city of Valletta (Figs. 3 and 4).

Fig. 3 View of the Grand Harbour from Valletta Upper Barrakka Gardens, in Malta

Fig. 4 Upper Barrakka Gardens—one of the most visited sites by tourists in Valletta

Before carrying out the requirements phase we met with experts in World War II history in Malta, as they gave us their accounts of the arrival of the SS OHIO. Further to this, we arrived to a decision that the moment that needed to be captured and brought to AR, rather than being something dramatic, had to capture the emotions of the people watching the merchant ship being towed in, as they sang, clapped and cheered—an emotion that was not totally understood even by the crew of the ship at that time. Thus we wanted to evoke in users that sense of empathy with the Maltese people who were waiting with hope for the last remaining ship of the convoy to make its way into the port. During the requirements phase it was decided that the most accessible way of disseminating the experience was through the use of Mobile AR (mAR). Therefore, the design factors had to take into account that the primary use of the app was for the mobile device. One of the first questions that we asked was, how do we associate and organise the information displayed in AR that can present the users with an experience that is not too lengthy but which conveys a specific moment in the history of Malta? The requirements stage helped us identify a number of design considerations for the app:

- Target Audience: The target audience was identified as primarily being tourists. Typically, tourists visit the location as part of a longer tour of Valletta. They stop at the gardens for a rest, and to enjoy the view from the bastions on to the Grand Harbour. Due to the large number of people taking photos and enjoying the view, and because usually this is not a long stop, tourists do not usually have much time to spare. This gave us an indication that the app experience needed to be short, yet convey an emotionally engaging experience.

- Location: The Barrakka Gardens was the identified location due to its public accessibility and space, and because it is one of the landmarks at Valletta. Since the gardens themselves are quite extensive, we needed to find an unobtrusive way to give the target audience the right directions via their mAR experience.
- Information: There are plenty of written records and historical documents regarding the state of Malta during the World War II and many refer specifically to Operation Pedestal. However, bearing in mind the target audience we decided to give out short, interesting snippets of information to stimulate curiosity about the event rather than give a full account of the whole event.
- User Interface: Using design considerations for mAR, using simple tactile-visual clues with graphics that are congruent with the era in history [7].
- Technology: Due to the nature of the situated simulation, the technology requirements included the use of marker-less AR, acquiring location coordinates data from the mobile device to trigger the AR simulation. Following Liestol's [15] design framework we chose to go for a layered approach in the identifying menu, where users could choose to either read through some of the historical records of the event or else disregard and start the simulation. Another design consideration for mAR is that user manipulation of the smartphone is quite limited—in a situated simulation, users may be moving as they view the AR from different angles. Thus, one has to plan for use where it is required that the user would hold the device with one hand, whilst moving or interacting with the app. This means that the amount of interactions may be limited and therefore we needed to provide support as much as possible in terms of low physical effort and the response of the application itself [3].

The small size of the display on users' smartphones increases the complexity of the projection of the graphics and details. The screen display needs to contain, enough information to aid navigation and access in as an intuitive way as possible whilst at the same time, leaving enough empty space to view the real world and overlaid digital augmentation without creating clutter. In design terms, this avoids confusion whilst at the same time, keeping all the information under the control of the user.

5.1 Design Approach: Theory Versus Practice

We approached the design phase by taking into account different considerations. The first consideration was that the scene had to be augmented over water within the harbour. This poses a number of challenges including changes in the weather, sea surface conditions as well as traffic within the harbour area.

The first possible approach that we took into account was that of using image recognition of the forts overlooking the harbour. The Grand Harbour in Valletta, is a site that is usually busy with cruise liners and other sea vehicles. Deeper research into the possibility made us reconsider the use of object recognition due to the position of physical ships that might obstruct the scene of the forts. Being an outdoor location, we noted that AR might also be affected by changes in the weather conditions. We also made considerations about using either ARKit or ARCore. The former, is for development for Apple OS whilst the latter, at the time of development, could only be used on new Android OS. Since our target audience was mainly tourists, we could not assume that people using the app would be carrying brand new Android devices and therefore we decided to avoid using this development kit.

Next we tested other different AR development systems, including Vuforia, Kudan and Wikitude. The challenges which we encountered when using these systems was always related to the location and the physical site. An indoor site or one which is not subject to continuous changes such as a busy harbour or port, presents a different scenario altogether which can utilise the technologies mentioned above. Vuforia and Wikitude, both presented problems related to the 2D plane as the sea is not recognised as a 2D plane. In addition, real ships entering or leaving the harbour area are recognised as foreign objects and this disrupts the whole AR experience.

Finally we chose to work with coordinates obtained from the user's smartphone device by utilising the inbuilt GPS tracker. We worked using values relative to the phone's orientation within a specified range. We specified a range of latitudes, longitudes and rotation—thus collecting three values. Whilst the geo-location values are continuously being updated, the user is being presented with the real-time environment, with the digital representation of the entry of the SS Ohio to the harbour overlaid on the users' screen. We used Unity [28] to be able to bring the AR experience to the audience's mobile devices. Having chosen the technology that offers the best affordances to work with, taking into consideration the external factors and environmental variables, we proceeded to develop the design approach user experience framework obtained from the research into mobile augmented reality and situated simulation applications.

Context provides content: AR-based research indicates that digital artefacts framed within a specific context provide for more meaningful experiences. Cultural heritage, particularly the sub-domain that focuses on historical events, derives value from the narratives that surround a specific phase. For the SitCity framework, for a situated simulation of a Maltese World War II historic event, we propose that context is key to the AR experience, and thus all the user-centric design components are described within the context of the setting.

Task relevant content: Information is presented in the initial home screen prior to the triggering of the simulation (see Figs. 5 and 6). In this way, users may choose to either read the information pertinent to the arrival of the SS OHIO in the Grand Harbour, or else start the simulation. This choice is also seen in Liestol's [15] situated simulation applications where users may choose whether to continue the simulation or else read more about the event or artefact. The scope of this component is to transfer the relevant information to the user in context (Fig. 7).

Fig. 5 Start menu for the mAR app with background information

Fig. 6 Background information about the SS OHIO

Provision of feedback: In SitCity users are required to be on location but for the geo-location aspect to trigger the simulation, they have to be within a specific range of co-ordinates. For this reason, the system is designed to give feedback to the users to be able to get within the right co-ordinate range for the simulation. We pay attention to the aesthetics presentation and the look and feel of the app, to make it as engaging as possible whilst offering the user with an enriched augmented experience (Fig. 8).

Usability: Usability is about offering the users a simple yet engaging experience. Users are presented with a screen that has textual notifications as well as arrows/icons that are familiar thus ensuring a more intuitive interface. Buttons are also kept at a

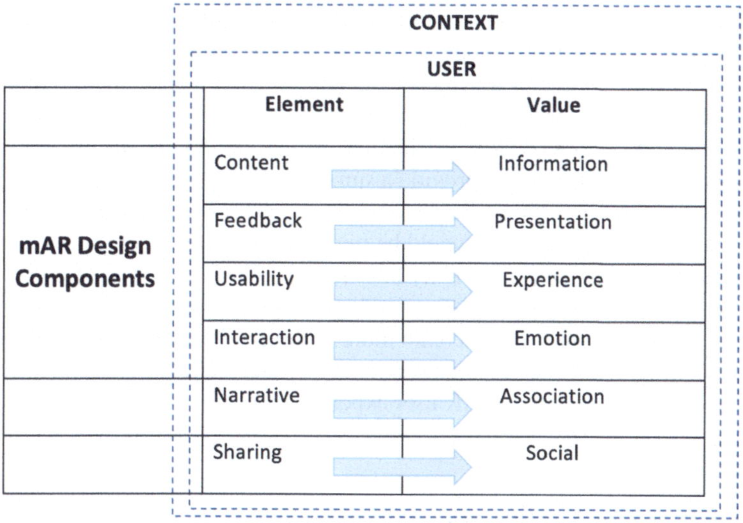

Fig. 7 Framework for the design of the user experience for SitCity

Fig. 8 Users are notified about their current position

familiar level, and are left as icons, so that it is easier for users to hold the mobile device with one hand and navigate the scenes with another. This semantic memory support design factor is utilised to keep the audience more engaged, whilst reducing the amount of frustration that would be involved in steep learning curves.

Interaction: Interaction is about creating an emotion with the appusers are required to be on location but for the geo-location aspect to trigger the simulation, they have to be within a specific range of co-ordinates.

Narrative: The story does not present any discourse. The narrative develops around the entry of the SS OHIO merchant ship as it is towed inside the Grand Harbour in Valletta. We chose not to present any documentary-style discourse to present the AR experience, but instead keep the realistic setting, by including the cheers, the sounds of jubilation and singing of "Rule, Britannia" as reported by the historical records. In addition, we have decided to provide a summary of the actual time during which the most famous ship of the convoy was being towed inside the harbour. The real event took hours to unfold. This would be impractical in the AR experience.

Sharing: Currently the pilot SitCity does not allow any user sharing options due to the fact that currently the narrative is still quite simple and the scenes shown are also limited and limiting. As we create more complex simulations, we believe that enabling sharing and some personalisation of the interactive components, can lead to a more engaging, retentive mAR user experience.

5.2 Testing and Evaluation

Working with situated simulations on location, especially if that location is an outdoors one, and if it includes moving surfaces like the sea, there are a number of challenges which are not easy to overcome.

The first challenge we encountered was lighting. The AR may experience some challenges due to the changing light variables, such as sunset, or a cloudy grey day as opposed to a sunshine filled day. External variables such as the weather impinge on object identification—so geo-location based AR, using surrounding buildings as trackers, may not be effective due to changing weather conditions.

The second challenge was related to the plane for the simulation. In this situated simulation, we present a scene whereby a merchant ship is being towed across the Harbour. Therefore, the plane of the simulation is the moving sea surface. Many AR Toolkits, such as Vuforia and Wikitude, do not recognise the sea as a 2D plane and therefore the AR generation is impeded.

There was also an added challenge related to the specific location chosen in Malta. Unity as an AR development platform, has an inbuilt location variable that has a default accuracy of 10m. Although the area of the public gardens is sizable, the part overlooking the Grand Harbour which is optimal for viewing the simulation offers a rather narrow passageway with a limited location span of movement, thus limiting the accuracy of the GPS latitude and longitude values.

Testing of the situated simulation was carried out to determine the effectiveness of the marker-less AR, and whether the users' geo-location coordinates are calculated in real time, thus refreshing the situated simulation seamlessly whilst the users are moving. The testing process involved unit testing through the Unity platform and an exploratory User Interface (UI) testing on location with 5 chosen users. Since there are no standard tests for AR apps we carried out the following tests to explore the user interactions with the app:

Location: If the user's mobile device is located too far from the railing overlooking the Grand Harbour, the user is directed to move forward/closer, and within the pre-set co-ordinates range to start the simulation. When an attempt is made to launch the app in an entirely different location, overlooking a different part of the sea (not the Grand Harbour), the application does not start the simulation.

Interaction: If the user is moving outside of the co-ordinates range, the phone's rotation is listed as not being within the range of pre-set co-ordinates, and therefore the user is prompted to redirect his mobile device back to facing the Grand Harbour. The user is shown arrows indicated the direction in which he/she has to turn the mobile device. As soon as the device is back in location the simulation starts. The user has the button options to go back to the simulation, read more information or move to the next scene.

AR scene: User testing carried out on location served to assess the realism of the situated simulation scenario. Although the assets at our disposal are rather limited, the digitally rendered ships are within the realistic dimension set. Smoke is added to represent the embattled state of the ship as it was towed in. As mentioned previously, rather than discourse we implement the sights and sounds of people cheering, as well as the band playing the "Rule Britannia". This helps to achieve a different form of experience and more in line with what was historically recorded.

5.2.1 User Evaluation

Feedback received from users on site provided us with a number of strengths and limitations and indications for future improvements of this first situated simulation prototype. Users reacted positively to the location chosen for the situated simulation. As detailed in the design approach, the mAR app was framed within the context of the simulation, therefore the location plays a very important role for the cognitive and affective transfer of the experience. The difficulties with the chosen location arise when the weather conditions are unfavourable, that is when it is too windy or rainy to visit the gardens and the balcony hanging out over the Harbour area, and when the area within the co-ordinates range is busy with other visitors viewing the panorama, thus limiting range of motion or securing the best place to view the simulation.

The design followed a user-centric approach, meaning that there is an emphasis on the role of the user as a consumer in view of all the design components that have been included. This includes the information relayed in the content, the presentation of the feedback given to the users, the overall experience impinging on the usability, the interactions and the narrative which contribute to enriching the affective experience. In terms of a user-centric approach, the content is presented as a short chunk of historical records outlining the entry of the SS OHIO in the Grand Harbour. However, from feedback received, some users requested additional information that included a broader perspective on Operation Pedestal, as well as any information about the other ships of the convoy, that were sunk and those that made it to safety. These can be inserted as additional info which users can choose to read through or else skip depending on their time availability and interest.

Users reacted well to the feedback displayed on screen especially when they moved outside the co-ordinates range and were redirected back within the range. In line with user-centric design, we make use of directional arrows as well as a short text indicating where best to locate to view the AR simulation. In our approach, we kept in mind, that the target audience of foreign tourists might have limited time and therefore they cannot use the time trying to learn to navigate through the app. Therefore, the feedback has to be brief, effective and help the users manage their time efficiently. We also took into consideration screen size, and whereas the target age of the audience was quite varied, we struck a balance between the size of the arrows and text, and their intrusiveness on screen.

This also leads to the usability factor of the overall app. SitCity UI design allows for a simple, easy to use and intuitive app, that users can download and access via their personal mobile devices. We also assume that users are familiar with handling their personal devices. The interface is such that it is as easy as possible for them to hold the device with one hand, and possibly interact with it using the other. For user interactions, we included buttons on the home screen and at the top of the screen during the simulation with obvious indications for the users to press. We also included text on the buttons for more accessibility.

The narrative has been kept short, and we avoid a documentary-style voice over, due to the fact that the target audience is expected to be outdoors and in close proximity to each other. With a voice over, users would be required to keep headphones, or the mobile speakers close to be able to hear over the din of the outdoors and crowd chattering. Therefore, whilst limiting the narrative to the sight of the SS OHIO being towed in, we added the cheering crowds and the music that was being played that day. Users reacted positively to this even though in their opinion the situated simulation was too short and they would have preferred additional AR scenes playing different World War II events happening in the Grand Harbour Area. This would be more effective to illustrate history in an alternative way.

One limitation of SitCity involves the AR assets. Although the ships and the crowds are to scale, the user can only see one perspective of the ship, whilst the ship itself and the smoke are not very realistic. The pilot currently does not allow for the ship to be rotated and viewed from different angles but this can be added as part of increased interactivity for future developments. Users may be able to select the ship, view the 3D model, and rotate it to view it from the different angles and perspectives.

User feedback indicated that future iterations of the mAR project should include the social aspect, which adds value to the overall AR experience. Currently the mAR app does not allow the users to either share their thoughts and pictures over social media. The situated simulation pilot, SitCity, has currently no further scope to broaden the educational and communication value. We believe that the mAR app can increase its value when sharing and creativity are added. As Jenkins et al. [10] imply, adding a participatory element would greatly benefit the social value of future iterations of the app or other additional mAR apps for museums and heritage sites. Using technology as a mediator for the cultural heritage field demands innovative uses that go beyond the mere application of the technology itself, but one which renders the users more active and participative in the experience and involves them as the medium through which cultural heritage is spread across networks.

6 Conclusion

This chapter illustrates how mobile augmented reality applications (mAR) can be used within the cultural heritage field. Although AR has been in use and popularised for some time due to the game Pokémon, as well as other applications for health, marketing and tourism, there is a new recognised need to bring this technology forward in the cultural heritage field. Heritage institutions can exploit AR technology, not just to assert their relevance in today's digital world but also because it can be used to convey a number of values, which can enrich an audience's experience, beyond the one visit to a site. We review research work in the field and discuss a number of design principles for mobile AR. One common design principle for mAR is the context in which these apps are viewed and interacted with. mAR apps are also characterised by task-related content presentation, real time feedback, usability and interaction.

We focus on a specific type of mAR, that uses geo-location to create situated simulations of an event in history. We illustrate a case example of a simulation that involves an event in World War II History of the Maltese Islands. We discuss the SitCity design factors involved in relation to the theoretical framework for the design of mAR apps, and illustrate the testing and evaluation of the pilot app. The theoretical implications are also discussed in terms of future research and development of this and similar types of mAR situated simulations. Although much of the existing research on AR and mAR focuses on the technology, usability and performance, we posit that for enhanced user engagement, the design elements need to be enclosed within the contextual framework. When the situated simulation is experienced within a defined context, a user can create meaningful experiences that are grounded into realistic settings. The domain of cultural heritage primarily deals with the identity of society irrespective of whether we view the tangible or intangible aspect of it. There is scope for further advancements in the research of mAR for cultural heritage to include the design and development of elements that enhance its social value. Through sharing, communication and co-creation, and through the application of technology, users can form part of a participatory culture that is not just responsive and sensitised to the subject but they can be a channel of transmission for spreading memorable experiences that can invoke further curiosity and generate interest.

Acknowledgements We would like to acknowledge the project work of the team behind SitCity, Anne-Marie Camilleri, Lara Caruana Montaldo and Lizzy Farrugia. We would also like to acknowledge the expert team, Matthew Camilleri and Dylan Seychell, from Colour My Travel, https://www.colourmytravel.com, for their on-site assistance.

References

1. Artivive—Bringing art to life (2018). https://artivive.com. Last accessed Oct 2018
2. E. Berzen, Use of augmented reality to increase learning in museums (Malmo universitet/Kultur och samhalle, Malmo, 2018)

3. J.D. Bolter, M. Engberg, B. MacIntyre, Media studies, mobile augmented reality, and interaction design. Interactions **20**(1), 36–45 (2013). https://doi.org/10.1145/2405716.2405726
4. A. Bonasio, *Augmented Reality Will Reinvent How We Experience Art* (2018). https://www.forbes.com/sites/alicebonasio/2018/06/20/augmented-reality-will-reinvent-how-we-experience-art/. Last accessed May 2019
5. G. Casimiro, *Augmented Reality and the Dimensions of Sensibility* (2018). https://artivive.com/augmented-reality-and-the-dimensions-of-sensibility/. Last accessed Feb 2019
6. Y.L. Chang, H.T. Hou, C.Y. Pan, Y.T. Sung, K.E. Chang, Apply an augmented reality in a mobile guidance to increase sense of place for heritage places. J. Educ. Technol. Soc. **18**(2) (2015)
7. S. Ganapathy, Design guidelines for mobile augmented reality: user experience, in *Human Factors in Augmented Reality Environments*, ed. by W. Huang, L. Alem, M.A. Livingston (Springer, New York, 2013), pp. 165–180. https://doi.org/10.1007/978-1-4614-4205-97
8. G. Genette, *Narrative Discourse. An Essay on Method* (Cornell University Press, Ithaca, 1980)
9. IKEA (2019). https://highlights.ikea.com/2017/ikea-place/. Last accessed May 2019
10. H. Jenkins, S. Ford, J. Green, *Spreadable Media: Creating Value and Meaning in a Networked Culture* (NYU press, New York, 2018)
11. Knight Foundation (2018). https://knightfoundation.org/about. Last accessed Oct 2018
12. P.E. Kourouthanassis, C. Boletsis, G. Lekakos, Demystifying the design of mobile augmented reality applications. Multimed. Tools Appl. **74**(3), 1045–1066 (2015)
13. J. Lave, E. Wenger, *Situated Learning: Legitimate Peripheral Participation*, vol. 521423740 (Cambridge University Press, Cambridge, 1991)
14. F. Liarokapis, An augmented reality interface for visualizing and interacting with virtual content. Virtual Reality **11**(1), 23–43 (2007)
15. G. Liestol, 'Sequence & Access, Storytelling & Archive in Mobile Augmented Reality', in *Proceedings of the 23rd International Conference on Virtual System & Multimedia (VSMM), Dublin, Ireland (2017)* ed. by L. Goodman, A. Addison IEEEXplore Digital Library, pp. 1–7. https://doi.org/10.1109/VSMM.2017.8346305
16. M. Llewellyn-Jones, *Operation "Pedestal": Convoy to Malta, 11–15 Aug 1942* (Royal Navy, Great Britain, 2012)
17. L.F. Marques, J.A. Tenedorio, M. Burns, T. Romao, F. Birra, J. Marques, A. Pires, Cultural heritage 3D modelling and visualisation within an augmented reality environment, based on geographic information technologies and mobile platforms. Arch. City Environ. **11**(33), 117–136 (2017)
18. N. Moorhouse, T. Jung, Augmented reality to enhance the learning experience in cultural heritage tourism: an experiential learning cycle perspective. eReview Tour. Res. **8** (2017)
19. I. Pedersen, N. Gale, P. Mirza-Babaei, S. Reid, More than meets the eye: the benefits of augmented reality and holographic displays for digital cultural heritage. J. Comput. Cult. Herit. (JOCCH) **10**(2), 11 (2017)
20. M. Reunanen, L. Diaz, T. Horttana, A holistic user-centered approach to immersive digital cultural heritage installations: case Vrouw Maria. J. Comput. Cult. Herit. (JOCCH) **7**(4), 24 (2015)
21. B. Ridel, P. Reuter, J. Laviole, N. Mellado, N. Couture, X. Granier, The revealing flashlight: interactive spatial augmented reality for detail exploration of cultural heritage artifacts. J. Comput. Cult. Herit. (JOCCH) **7**(2), 6 (2014)
22. Y. Rogers, H. Sharp, J. Preece, *Interaction Design: Beyond Human-Computer Interaction* (Wiley, New York, 2011)
23. Smithsonian National Museum of Natural History: A Hall through New Eyes (2018). https://naturalhistory.si.edu/exhibits/bone-hall/. Last accessed Oct 2018
24. O. Smordal, G. Liestol, O. Erstad, Exploring situated knowledge building using mobile augmented reality. QWERTY **11**(1), 26–43 (2016)
25. F. Tscheu, D. Buhalis, Augmented reality at cultural heritage sites, in *Information and Communication Technologies in Tourism 2016* (Springer, Cham, 2016), pp. 607–619

26. M.C. Tom Dieck, T.H. Jung, Value of augmented reality at cultural heritage sites: a stakeholder approach. J. Destin. Mark. Manag. **6**(2), 110–117 (2017)
27. M.C. Tom Dieck, T. Jung, A theoretical model of mobile augmented reality acceptance in urban heritage tourism. Curr. Issues Tour. **21**(2), 154–174 (2018)
28. Unity (2019). https://unity.com. Last accessed June 2019
29. M. Van Der Vaart, *Using Augmented Reality in the Museum: Material Encounters with Digital Cultural Heritage* (2014). http://mesch-project.eu/using-augmentedreality-in-the-museum. Last accessed Feb 2019
30. Virtual Reality Pop (2019). https://virtualrealitypop.com/aredu-educational-augmented-reality-apps-5e6599529807. Last accessed May 2019
31. J. Westin, A. Foka, A. Chapman, Humanising places: exposing histories of the disenfranchised through augmented reality. Int. J. Herit. Stud. **24**(3), 283–286 (2018)
32. Within (VR) (2018). http://milk.co/vr. Last accessed Oct 2018

Glossary

Artificial Intelligence The creation of computer programs and machines that can replicate human intelligence. AI includes learning from information, reasoning about the resultant rules and self-correction given all of the above.

Artificial Neural Networks An ANN machine learning approach inspired by the structure of the human brain. ANNs are built by several components, called neurones, whose weights are adjusted depending on the relationship between the input and the desired output. The more examples fed into the ANN, the better the learning and the better the success of the network.

Augmented Reality A technology that superimposes a computer-generated image on a user's real view of the real world captured through a camera, thus providing a composite view. AR can be experienced through head-mounted displays, handheld devices or spatial configurations.

Computer Vision An interdisciplinary scientific field that deals with how computers can be made to gain a high-level understanding from digital images or videos. This discipline includes image acquisition, image processing and image understanding through AI techniques.

Deep Learning A branch of machine learning that makes use of more complex ANN architectures resulting in a larger number of hidden layers of neurons, thus making the network deeper. This approach has been proved to be very successful in performing narrow tasks, given enough data.

Gamification The application of game-design elements and game principles in standard routines that are normally performed in a more serious environment. The allows the users to learn more about the specific domain within a safe environment of a game that would also track their progress and allow for improvement in specific tasks.

Learning Strategies This refers to procedures and strategies that learners use to gain further knowledge and skills in a particular domain. Such strategies are used to assist learners, depending on their different capabilities.

D. Seychell and A. Dingli (eds.), *Rediscovering Heritage Through Technology*, Studies in Computational Intelligence 859,
https://doi.org/10.1007/978-3-030-36107-5

Machine Learning A major field of research within Artificial Intelligence that uses algorithms and statistics to make sense out of data without a predefined set of rules. This approach relies on the patterns discovered within the data and the respective inferences.

Natural Language Processing A subfield of artificial intelligence concerned with the interactions between computers and human languages. This discipline focuses on how to program computers to process and analyse natural language data and extract semantic information from it.

Systems Analysis The activity of collecting information about a procedure or set of processes. This is carried out to identify the purpose of a system that eventually allows for the development of technological solutions that enhance the current system.

User Modelling The process of building a conceptual model that represents the attributes and behaviour of a user. This approach is used to make the software more adaptable to different user needs to deliver tailored solutions and results.

Virtual Reality VR delivers an experience taking place within a simulation, which can be similar to or completely different from the real world. Users experience VR by wearing a head-mounted display that provides a dynamic view of the virtual world and adjusts itself to the movements of the user.

Index

D. Seychell and A. Dingli (eds.), *Rediscovering Heritage Through Technology*, Studies in Computational Intelligence 859,
https://doi.org/10.1007/978-3-030-36107-5

The manufacturer's authorised representative in the EU is Springer Nature Customer Service Centre GmbH, Europaplatz 3, 69115 Heidelberg, Germany. If you have any concerns regarding our products, please contact ProductSafety@springernature.com

Printed and bound by CPI Group (UK) Ltd, Croydon, CR0 4YY
05/10/2026
02228120-0004